Paul Adrien Maurice Dirac, one of the greatest physicists of the twentieth century, died in 1984. Dirac's College, St John's of Cambridge, generously endowed annual lectures to be held at Cambridge University in his memory.

This volume contains a much expanded version of the 1994 Dirac Lecture presented by Nobel Laureate Pierre Gilles de Gennes. The book presents an impressionistic tour of the physics of soft interfaces. Full of insight and interesting asides, it not only provides an accessible introduction to this topic, but also lays down many markers and signposts for interesting new research possibilities. The text begins with a brief discussion of wetting and dewetting and then goes on to consider the dynamics of different types of interface before considering adhesion and polymer/polymer welding.

Condensed matter physicists, material scientists, physical chemists and chemical engineers will find this book of interest.

Soft Interfaces

Other Dirac Lectures

Richard P. Feynman and Steven Weinberg
Elementary Particles and the Laws of Physics

A. Salam
Unification of Fundamental Forces

R. F. Dalitz and P. Goddard
Dirac: aspects of his life and work

Soft Interfaces

The 1994 Dirac Memorial Lecture

P. G. de Gennes

Collège de France, Paris

CAMBRIDGE
UNIVERSITY PRESS

CAMBRIDGE UNIVERSITY PRESS
Cambridge, New York, Melbourne, Madrid, Cape Town, Singapore, São Paulo

Cambridge University Press
The Edinburgh Building, Cambridge CB2 2RU, UK

Published in the United States of America by Cambridge University Press, New York

www.cambridge.org
Information on this title: www.cambridge.org/9780521564175

First published 1997
This digitally printed first paperback version 2005

A catalogue record for this publication is available from the British Library

ISBN-13 978-0-521-56417-5 hardback
ISBN-10 0-521-56417-4 hardback

ISBN-13 978-0-521-02035-0 paperback
ISBN-10 0-521-02035-2 paperback

Contents

Contents

Introduction

Dirac was a man who concentrated on the difficult problems of his time. He was principally interested in the basis of quantum mechanics and the elementary particles. However, on one occasion, around 1938, he did write a paper which went to the opposite extreme and discussed the size of the cosmos and the age of the universe in terms of very simple dimensional analysis; such data is still alive and well and still food for thought.

As the centuries have gone by, physicists have of course tended always to move in those directions where great problems remain, and I think if one looks at the progress in physics until, shall we say, the 1940s, they have definitely concentrated on very small things — atoms and small molecules. In the period, in the fifties, then the sixties, it was realised that the methods of physics could be applied to other regions which lay above the atomic scale of, shall we say, everyday life — and by that I mean below the scale of what one normally thinks of as hydrodynamic phenomena. There was a mesoscopic physcs, an intermediate scale where the ideas and the methods of physics could work and make progress. One of the leading workers in this area, one of the prophets, is Pierre Gilles de Gennes. In the last few decades he has produced an enormous number of ideas which have been directly applicable to the experimental world, and which indeed have come to dominate our study of that world. It will be one of these areas that he will talk about here.

He is a man who has received many great honours, and perhaps to the outside world his Nobel Prize is the summit of these honours. I'm very proud to say that the University awarded him an honorary doctorate *before* he got the Nobel Prize, because any university can award one after someone gets the Nobel Prize. I think it is clear there was a proper appreciation here of his talents, and I am very pleased that the Dirac Lecture of 1994 is now to be given to us by Pierre Gilles de Gennes.

Professor Sir Sam Edwards

I
Geography and explorations

The borders between great empires are often populated by the most interesting ethnic groups. Similarly, the interfaces between two forms of bulk matter are responsible for some of the most unexpected actions. Of course, the border is sometimes frozen (the great Chinese wall). But in many areas, the overlap region is *mobile*, *diffuse*, and *active* (the Middle East border of the Roman empire; disputed states between Austria and the Russians, or the Italians, ...).

At a certain naive level, these distinctions can be transposed to physical interfaces between two different forms of matter.

(1) The hard frozen surfaces of metals, of ionic solids, or of semiconductors can be studied under conditions of high vacuum: this allows us to probe them — using electron beams, or other radiations which extract electrons from the surfaces; or even beams of neutral atoms. The net result is, in our days, a highly sophisticated knowledge of these sharp robust fortifications.

(2) The soft interfaces built from liquids, from polymers, from organic solids, or from detergents are much harder to probe. High vacuum is usually not acceptable. And even if it is, the probing beams can damage the interface. For many centuries, the main information on soft interfaces came from *mechanical* studies: adhesion, slippage, wear, ... During the last fifty years, *electrical* properties have also been helpful – in particular for the electric 'double layers' at the contact region between water and a solid.

More recently, a number of new tools became available:

(*a*) *Reflectance techniques*, using short-wavelength radiations such as X rays or neutrons.

(*b*) *Atomic force microscopes*, which can be used even in the presence of a liquid.

(*c*) *'Environmental' scanning electron microscopes*, which can operate under finite water pressures, allowing us to retain the native structure of wet surfaces. (I have just discovered this instrument here at the Cavendish.)

From the point of view of soft interface physics, the present times are fortunate: centuries of empirical knowledge about 'tribology' (friction) or 'colloids' (ultradivided matter) can progressively be correlated to detailed structural data at the 10 Å level. It is especially pleasant for me to mention this in Cambridge, where *the* major advances on tribology have been achieved — by F. P. Bowden, D. Tabor, and their co-workers. Of course, I shall not try to redescribe this sector. But I will insist on some general features of soft interfaces which were mentioned at the start: borders which can be *mobile*, *diffuse*, and *active*.

II

Mobile borders: the dynamics of wetting (or dewetting)

Since the days of Thomas Young, we know that a liquid (L), when deposited on a flat, impermeable, solid surface (S), may show two types of equilibrium behaviour: partial wetting (figure 1a) or total wetting (figure 1b). The choice is dictated by the interfacial energies $\gamma_{SL}, \gamma_{SA}, \gamma_{LA}$ (where A stands for the air[*]).

When the combination:

$$S = \gamma_{SA} - (\gamma_{SL} + \gamma_{LA}) \qquad (1)$$

is positive, the energy of the solid/air interface is lowered by intercalation of a flat liquid film: this corresponds to complete wetting. But when S is negative, a liquid drop does not spread on the solid: it terminates in the form of a wedge, with a well-defined contact angle θ_e (figure 1). We call this partial wetting.

---

[*] In the following text, we shall often use the shorthand $\gamma_{LA} = \gamma$ for the liquid surface tension.

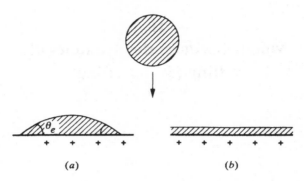

Mobile borders

(*a*)   (*b*)

Figure 1. (*a*) Partial wetting. (*b*) Total wetting.

Balancing the tensions γ (projected along the solid surface, which defines the allowed direction of motion), Young found the admirable relation:

$$\gamma_{SA} - \gamma_{SL} = \gamma_{LA} \cos\theta_e \qquad (2)$$

1 Dynamics of partial wetting

Equation (2) holds at equilibrium. What happens if we move out of equilibrium, for instance, by forcing a droplet on a surface, or by other experiments, displayed on figure 2? Let us discuss this for the case of partial wetting [1].

If the contact line of figure 3 moves at a velocity V, we expect a dissipation per unit length:

$$TS = FV \qquad (3)$$

6

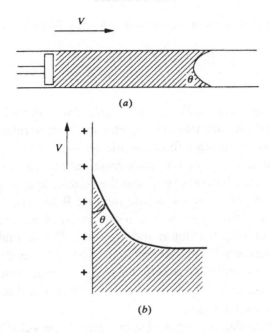

Figure 2. Schematics of two experiments to examine partial wetting.

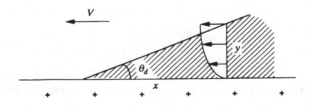

Figure 3. A wedge of liquid moving with velocity V.

where F is the non-compensated Young force:

$$F = \gamma_{SA} - \gamma_{SL} - \gamma_{LA}\cos\theta_d$$
$$= \gamma_{LA}(\cos\theta_e - \cos\theta_d) \qquad (4)$$

θ_d being the dynamic contact angle. If we can find the dissipation mechanism, we end up with a relation between the driving force and the velocity.

The dissipation may have different origins: either molecular processes very near the contact line; or viscous processes in the whole moving fluid. The first may be sensitive to the chemical details of the molecules making the liquid and the solid. The second is more universal. There is one limit, when viscous flows must be dominant: namely when the dynamic contact angle is small ($\theta_d \ll 1$). We can understand this by the following argument.

Inside the moving wedge of figure 3, the velocities v range from $v \sim V$ at the free surface and $v \sim 0$ at the lower surface. Therefore the viscous dissipation is of order:

$$T\dot{S} = \int dx\, \eta \left(\frac{V}{y}\right)^2 y \qquad (5)$$

where $y = \theta_d x$ is the local thickness. Equation (5) gives a logarithmic integral $l = \ln(x_{\max} / x_{\min})$. Putting in the correct coefficients:

$$T\dot{S} = 3l\frac{\eta V^2}{\theta_d} \qquad (6)$$

The logarithmic factor *l* is typically of order 12; it has worried the experts in fluid mechanics for many years. But it is not the dominant feature of equation (6): the really important feature is the presence of θ_d in the denominator. At small wedge angles, the viscous dissipation becomes very large, and dominates over all molecular processes.

A careful reader may object to this simple discussion, since it contains a hidden assumption: the moving liquid profile near *L* is taken to be still a simple wedge ($y = \theta_d x$). Could it, in fact, be more singular? We know the answer from singular perturbation calculations of Cox [2], or from simpler methods. There are indeed corrections to the simple wedge, of the form:

$$y = \theta_d\, x\left(1 + k\frac{V\eta}{\gamma}\ln\frac{x}{x_{\min}}\right) \qquad (7)$$

(where *k* is a numerical constant, and $x_{\min}$ depends on molecular features or on the presence of long-range forces [3]). The crucial parameter here is the capillary number $Ca = V\eta/\gamma = V/V^*$. In practice, the velocities *V* of interest turn out to be always of order:

$$V \sim V^*\theta_d^3 \ll V^* \qquad (8)$$

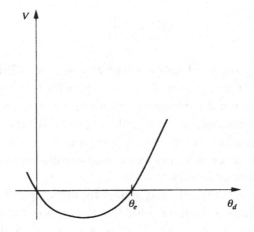

Figure 4. Dynamic contact angle θ versus velocity V.

and the corrections to the profile are thus not very important for the discussion of $F(V)$. But they have been seen in direct optical observations by E. Dussan and coworkers [4].

Returning now to equations (3) and (6), we end up with a basic dynamic formula (for partial wetting):

$$F \equiv \gamma(\cos\theta_e - \cos\theta_d) = 3l\eta V / \theta_d \qquad (9)$$

valid for $\theta_d \ll 1$. A vast number of experiments (some of them depicted in figure 2) can be understood simply in these terms [1].

Figure 4 shows the relation between V and θ_d in partial wetting. Of course, V vanishes at the equilibrium angle $V(\theta_e) = 0$. But V also vanishes at small θ, where the dissipation is large.

2 Complete wetting

For complete wetting, the driving force is (always from the Young argument):

$$F = S + \gamma(1 - \cos\theta_d) \qquad (10)$$

In the most interesting limit ($\theta_d << 1$), this force is nearly constant:

$$F = S + \tfrac{1}{2}\gamma_{LA}\theta_d^2 \cong S \qquad (11)$$

Experimentally, a number of experiments on spreading droplets, or on liquids pushed in wettable tubes, show that the resulting velocity V is in fact independent of S! The observed spreading law is:

$$V = (\text{const})V^*\theta_d^3 \qquad (12)$$

The explanation for this anomalous behaviour [5] is based on the existence of a *precursor* film (figure 5) first observed in elegant experiments of Hardy (1919). The free energy described by S is 'burned' via high shear flow inside the precursor film: the macroscopic wedge is driven only by the residual force $F - S = \tfrac{1}{2}\gamma\theta_d^2$. Using this modified form, plus the dissipation formula (6), one easily reaches equation (12).

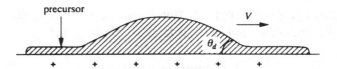

Figure 5. The precursor film first observed by Hardy (1919).

3 Molecular features

All the previous discussion was essentially macroscopic (except for the inner structure of the precursor film, where long-range Van der Waals attractions from the solid play a leading role). Molecular features can, in fact, show up under various guises.

(*a*) At large dynamic angles θ_d, the hydrodynamic losses do not necessarily dominate. Molecular processes, very near the contact line, can become important. One of these is shown on figure 6, where a molecule from the liquid hops through the vapour phase, and hits the adjacent solid surface. The crucial parameter here is clearly the energy of vaporisation E_v, a plausible relation between F and V being now of the form:

$$V = kF \exp\left(\frac{-E_v}{kT}\right) \qquad (13)$$

We do not have, at the moment, a full estimate of the prefactor $k(\theta_d)$.

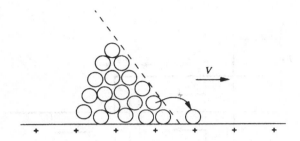

Figure 6. Molecular processes near the contact line.

(*b*) When very thin droplets (or precursor films) are observed, *terraced structures* often show up — especially with liquid molecules which are more-or-less spherical [6]. The first terrace (one monolayer of liquid on the solid surface) can exist in various forms: as a two-dimensional gas, or as a two-dimensional liquid (and a coexistence line separating the two forms can also show up). Here, let us restrict attention to cases where each layer behaves like an incompressible liquid.

The two basic processes involved in the dynamics of a terraced droplet are shown on figure 7.

The first, of course, is *friction* between adjacent layers, which move at different velocities. The second is *permeation*, where molecules move from one layer to the next, the flow rate being proportional to the local difference in chemical potential. When these two ingredients are put in [7], one finds that permeation

13

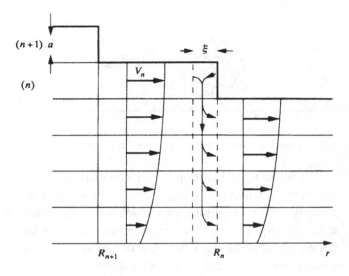

Figure 7. Dynamics of a terraced droplet.

occurs only in a thin ribbon, of microscopic width ξ, near each step (figure 7). Ultimately, for terraces which are much larger than ξ, one can replace the detailed permeation process by a boundary condition: near each step, the chemical potentials of the two relevant layers are equal.

For a two-layer system (with radius R_1, R_2) the bottom layer spreads out ($R_1 > 0$) while the top layer shrinks ($\dot{R}_2 < 0$) by permeation. The inner part of both layers ($r < R_2 - \xi$) is predicted to be completely static! We have no direct experimental proof of this.

14

But some verifications of the growth laws have been obtained on a short cyclic siloxane ('tetrakis') which is a spheroidal molecule [8]. There is one complication, however: the density of tetrakis in the first layer is not quite constant − we are really dealing with a compressible two-dimensional fluid.

When there are more than two terraces, many dynamical behaviours are possible, depending on the initial conditions, and also on the detailed form of the long-range energies W_n between the solid and the nth layer. This represents an expanding, amusing domain, both for experiments and simulations.

III

Decorated borders: slippage between a solid and a polymer melt

I like to work more with examples than with general statements. Typical borders of interest are depicted on figure 8 overleaf:

(*a*) a polymer melt facing a simple solid;

(*b*) a solid decorated by grafted chains;

(*c*) the bare interface between two polymer melts, A and B, which is also somewhat diffuse if A and B are similar chemically;

(*d*) the intelligent animal which we call a block co-polymer, intercalated between two polymer melts.

Figure 9 shows another interesting situation, with two networks, or rubbers, facing each other. Added to this are some mobile chains, which wander around. At some moment, they may provide a *transient bridge* between the two sides. This is sometimes of interest in the rubber industry, when you want to glue two pieces together (free chains are naturally present in weakly vulcanised rubber). It is also useful at a much smaller scale, when you have latex particles and want them to fuse, generating a protective coating.

Decorated borders

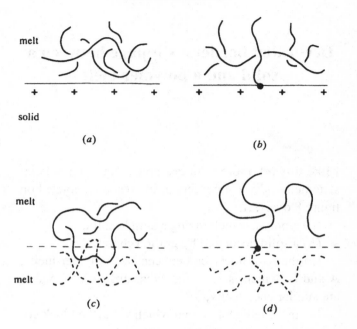

Figure 8. Borders: (*a*) between a polymer melt and a simple solid; (*b*) a solid decorated by grafted chains; (*c*) a bare interface between two polymer melts; (*d*) a block copolymer intercalated between two polymer melts.

Decorated borders

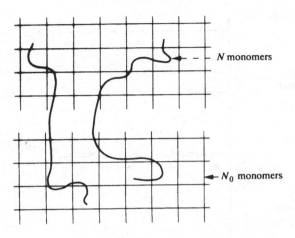

Figure 9. Two networks, or rubbers, facing each other.

Let me return to the stupid solid facing a polymer melt (figure 10) and discuss slippage. This would seem to be a simple situation, but it is not really. In our youth, we studied classical mechanics (in the rather strong and firm language which this community uses), and we have been taught that velocity fields are *continuous at a boundary*. If the solid is at zero velocity and if we impose some flow in the fluid above, the velocity of the fluid should vanish at the interface.

Now, of course, we should take this with a grain of salt. If we think of simple liquids at the molecular level, we realise that this boundary is not really sharp: it is fuzzy, at least at the scale of the atoms or molecules which make up the liquid. We therefore suspect

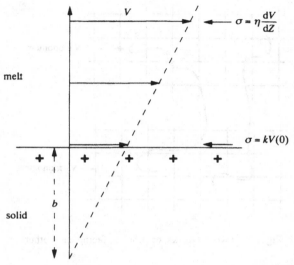

Figure 10. A stupid solid facing a polymer melt.

that there could exist a small extrapolation length. Indeed, you can argue, as shown on figure 10, that the extrapolation length b is a ratio of the viscosity of the liquid η to the friction coefficient of the surface k:

$$b = \frac{\eta}{k} \tag{14}$$

where k is defined by the relation $\sigma = kV(0)$ between stress and surface velocity. For a simple liquid, the length b will be comparable to a molecular size a (a

20

few angstroms) and we will not be able to see it in most experiments.

However, if you turn to the polymer case, and assume that the polymer melt is facing a passive surface, you realise that things could be very different. What I mean by a passive surface is one with certain properties: no chemical bonds with the surface; a smooth interface; no strong Van der Waals interactions which could create a glassy layer of the liquid near the surface.

The polymer liquid has an enormous viscosity (because it is an entangled system), but it need not have a large friction coefficient: the wall friction acting on any unit in a polymer chain is still exactly the same as it would be in the corresponding simple liquid of monomers. If a is the size of a monomer unit, then:

$$k = \frac{\eta_{mono}}{a} \tag{15}$$

Thus, you finish up with an extrapolation length involving a and the ratio of the viscosities of the entangled melt and the monomer liquid:

$$b = a\frac{\eta_{melt}}{\eta_{mono}} \tag{16}$$

In the classic reptation model (in terms of polymer

length N and entanglement distance N_e measured as multiples of the basic monomer), the ratio is of the form:

$$\frac{\eta_{\text{melt}}}{\eta_{\text{mono}}} \sim \frac{N^3}{N_e^2} \tag{17}$$

The viscosity ratio is enormous: you can tune it by choosing the polymer length, and it can be in the range of 10^6 to 10^9. You are therefore tempted to think that these ideal passive surfaces should show an enormous slippage.

When I first stated this at a rheology meeting in Naples [9], it was received with mixed feelings – and for obvious reasons. Many in that audience had measured viscosities of polymer melts in capillaries for a long time, and my statement was casting doubt on these measurements. However, after some time, it became clear that in weak flows, under most practical conditions, when you use an arbitrary capillary (made of steel, or glass, or something similar) it is so dirty that the ideal surface property I invoked above does not apply. There are, nevertheless, a few cases where you do see slipping, and I have depicted three of the relevant experiments in figures 11 to 13 below. I like these from the point of view of the history of science, especially the first one.

Decorated borders

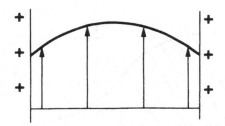

Figure 11. A schematic of Bryce
Maxwell's slippage experiment.

Bryce Maxwell, the famous professor at Princeton, was one of the founders of the science of plastics in the USA in the post-war years. He was interested in the way a polymer melt is 'extruded' (by a heavy mechanical system, pushing the molten polymer with a screw, under high pressure), and wanted to see what happened during this process. He was talented enough to build a transparent extruder, made out of glass but nevertheless resisting the pressures (400 atmospheres) that an extruder must produce. Once he had built this, he could look at the flows by adding little particules to label the trajectories. He found that in his transparent dye he very often had a finite slippage velocity (figure 11). Now Princeton is a rather well-known place for mechanics, so Bryce invited a number of his colleagues to observe the phenomenon. They refused to believe it. In fact, Bryce had great difficulty in publishing his paper. Ultimately, it came out in some

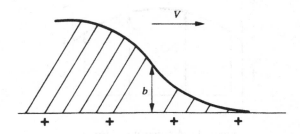

Figure 12. Slippage experiment using polymer drops.

obscure plastics journal, but it is an important piece of history.

The second type of experiment showing slippage is related to the spreading of drops. Certain polymer drops, when they spread on a solid surface, show an anomalous 'foot' (figure 12). Françoise Brochard has interpreted this foot as a consequence of slippage. These experiments are interesting because they are concerned with very low velocities (the drop spreads in about one week).

Similar (but more precise) information can be obtained in *dewetting* on very clean (silanised) surfaces (figure 13). In a typical dewetting experiment, a dry patch grows on the solid surface, and the liquid collects into a rim. If the height h of the rim is much larger than the length b, one expects the classical form of dewetting (described in the previous chapter) con-

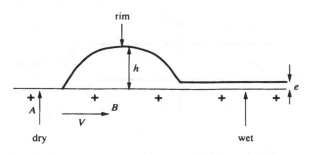

Figure 13. A dewetting experiment.

trolled by the losses at both edges of the rim, which are independent of the rim size. This leads to a constant growth velocity for the dry patch. On the other hand, if $h < b$, the flow in the rim is a plug flow, the total friction of this plug onto the solid is proportional to the width of the rim, and this increases with the amount of water collected: here, the dewetting velocity is not constant, but is a decreasing function of time. This has been observed recently by C. Redon and F. Brochard-Wyart [10], using high molecular weight silicones and very good surfaces, to reach a high b.

Finally, one example where plate rheometry was used, we owe to the British school (figure 14). You have polystyrene between two blocks of copper and you find that the apparent viscosity of the system decreases when the gap decreases. This greatly surprised Andrew Keller, who began doing this experiment for

25

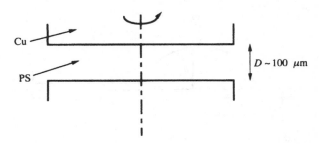

Figure 14. Slippage experiment using polystyrene between
two blocks of copper.

a completely different reason related to gelation in
polymer melts [11]. He thought that gels would build
up near the surfaces, and that if you pushed the plates
together, these gels would link together and there
would be an enormous viscosity. But no, it was
exactly the opposite! In fact, from this sort of experiment, you can deduce that for this particular system −
copper, polystyrene, with certain surface conditions −
you have a slippage length of 50 microns, which is
huge at the atomic scale[†].

These were the early days, but we now have a tool
which can tell us much more. It is based on an optical
technique. Suppose you have a polymer melt flowing
horizontally over a surface, as shown in figure 15a. In

[†] The surface velocities involved in this experiment are not small:
we may be dealing here with non-linear regimes, where the length
b is rapidly increasing with velocity (as we shall see later).

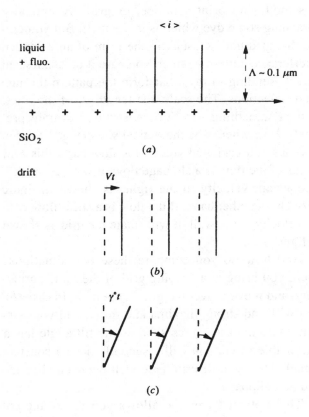

Figure 15. A schematic of a modern slippage experiment based on an optical technique.

this melt, you print a lattice, or grid, by optically bleaching some dye which is in the melt. But you create this grid not, as usual, in the form of an extended interference pattern – rather, you base it on evanescent waves, sending in rays that form the pattern through total reflections. The scale, the length Λ of the lines, is then something like 500 angstroms, or more precisely $\lambda/4\pi$ where λ is the optical wavelength. So you have a small grid, and you let the flow carry this grid along. If the flow is a slippage flow, it will just push it at constant velocity to the right, as shown in figure 15b. On the other hand, if the flow is a shear flow with no velocity at the wall, it will rotate the grid as shown in figure 15c.

And how do you compare these two situations? Well, you bring in a probing grid of the same periodicity, and move it over the grid in the melt. In case (b), you will find strong maxima and minima in your signal; while in case (c), as soon as the tilt angle has a reasonable value, you will essentially have a constant signal. Thus you discriminate well between slippage and non-slippage.

This optical technique allows you to do anemometry very close to a surface. Although classical laser anemometry works at 30 microns from a surface, this method works at a tenth of a micron from the surface. So it is a completely different ball park, and it has been very useful.

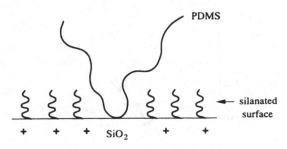

Decorated borders

PDMS

← silanated surface

SiO_2

Figure 16. A solid surface against a polymer fluid.

The system that L. Léger and her coworkers [12] looked at, using this tool, is depicted in figure 16. Here we have a solid surface against a polymer fluid (polydimethylsiloxane, or PDMS).

The surface is grafted and passive at most points, but there are a few holes. Some PDMS chains will attach permanently to these holes: we know from other experiments that siloxane binds to a silica surface, so you build up a surface which contains a few fixed chains, but not very many. In the jargon I introduced some time ago, we have a 'mushroom' regime, where there are only a few mushrooms scattered at different places – as opposed to a dense system, which we call a 'brush'.

This is the starting point of the investigation. Of course, when I say this, this is a typical theorist's statement – the preparation of these surfaces is not so simple, and the little drawing in figure 16 represents

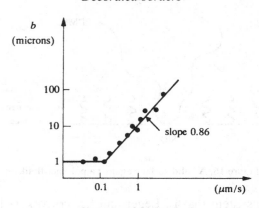

Figure 17. The extrapolation length, *b*, for
horizontal flows on the surface at various
velocities.

three years of work. However, once you have this, you
can look at the extrapolation length *b* for horizontal
flows on the surface at various velocities; an example
is shown on figure 17.

At very low velocities the extrapolation length is
small, which means no slippage. But when you go
beyond a certain threshold (which for this particular
example would be something like a tenth of a micron
per second) you find a progressive increase (by up to
four orders of magnitude) of the slippage length.
Thus, there is a slippage transition, but it is not an
abrupt jump; instead it is governed by a power law,
with an exponent of the order of 0.86 – possibly equal
to unity.

Decorated borders

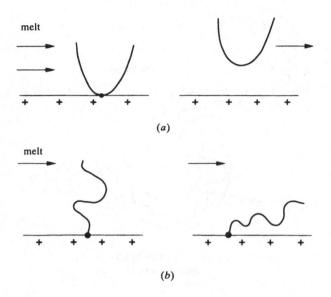

(a)

(b)

Figure 18. Possible interpretations of the slippage transition.

At first sight, there could be many interpretations of this observation. For instance, you might think of a simple tear-out of fixed chains (figure 18*a*). However, the tear-out explanation is not satisfactory because it would not give you the progressive increase. So, you are left with another explanation: at low velocities the chains involved are undisturbed mushrooms, and at high velocities they bend in the flow as in figure 18*b*.

When I was a child, I was taught some English, in

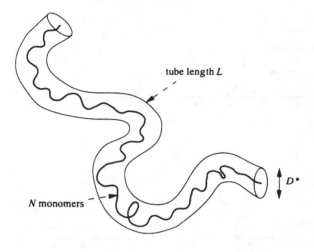

Figure 19. A chain in the melt behaves as if it were
trapped in a tube.

a book called *The Wind in the Willows**. Thus I like to
describe the situation by saying that if the wind is
mild, the willows are not too perturbed, but if the
wind is strong, the willows bend.

How does this work in more detail? Our starting
point will be again a Cambridge idea, which is very
important for the description of these entangled melts.
The idea (which comes from Sam Edwards) is that
each chain in the melt behaves very much as if it were

---

* I am deeply grateful to Prof. Barenblatt for providing me with a
copy of this charming book just after the talk.

trapped in a certain tube (figure 19). The diameter of the tube is related to what we call the entanglement distance, the distance between knots in the structure; it may be something like 50 Å in typical examples. Each chain is trapped in its own tube. Usually, a chain can escape by moving along the tube; however, this does not apply to the grafted chain, which is attached at one end. In fact, the only way for the grafted chain to escape is via a neighbouring chain moving out of its own tube in the same region: this will allow the necessary jump to take place (figure 20).

Thus, when the grafted chain (with N monomers) has moved relative to the melt by an amount D^* (the tube diameter), the melt chain (with P monomers) must have moved by a length comparable to its total tube length L_t (figure 19). This implies the tube velocity V_t of the (P) chain must be much larger than V:

$$\frac{V_t}{V} = \frac{L_t}{D^*} = \frac{P}{N_e}$$

The dissipation (associated with *one* (P) chain which is entangled with our grafted chain) is:

$$T\dot{S}_1 \cong \zeta_1 P V_t^2 = \zeta_1 \frac{P^3}{N_e^2} V^2 = \eta a V^2$$

where ζ_1 is a monomer friction coefficient, and η is

33

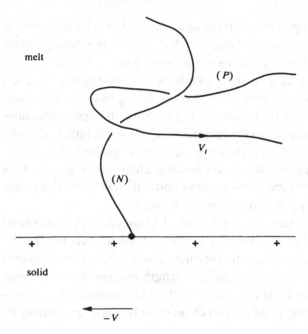

Figure 20. A grafted chain can only escape via a
neighbouring free chain.

the bulk viscosity of the melt (proportional to P^3 / N_e^3
in the reptation model).

The main problem is to find the 'drag number' X:
that is, how many mobile chains have to move out for
the willow to move. We essentially assumed in earlier
publications that all the chains which enter the willow's space participate. If N is the number of mono-

Decorated borders

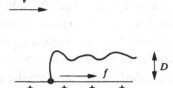

Figure 21. Strong deviation of a
polymer chain.

mers in the tethered chain, the region concerned has a
size $R = N^{1/2}a$, and a volume $N^{3/2}a^3$. Let us assume
$P > N$. Then each mobile chain entering the region
brings in a number of monomers $\sim R^2 / a^2 = N$. Thus:

$$X = N^{3/2} / N = N^{1/2}$$

The overall friction coefficient of the willow $X \sim \eta$
and is proportional to X, or to R. This is very similar
to Stoke's law for a sphere:

$$f = 6\pi\eta RV \qquad (18)$$

[However, this estimate of X may not be correct.
Other possibilities have been discussed recently by A.
Ajdari, F. Brochard-Wyart, C. Gay and J. L. Viovy:
see the Appendix. Here I shall stick to the simple
guess of equation (18).]

Given equation (18), it is easy to ascertain at which
moment the willow will begin to suffer (figure 21). In

polymer science, we know that if a chain is pulled with a certain force f, the chain begins to suffer strongly — to deviate from its original equilibrium conformation — when the force is something like kT, the thermal unit, divided by the unperturbed size; or in other words:

$$f = \frac{kT}{R} \qquad (19)$$

This is a pretty general theorem, independent of the detailed statistics of this chain — it may be an ideal object, or it may be a complicated interacting object. Plugging this into the friction formula at low velocities gives an estimate of the threshold velocity for the problem:

$$V^* \sim \frac{kT}{\eta R^2} \qquad (20)$$

This is the velocity at which the willow will begin to suffer. The numbers one obtains are very small (because of the melt viscosity in the denominator); they are in the range seen by Miegler, Léger and Hervet [12]. Recently, this group found that $V^* \sim N^{-1}$ as expected from equation (20) [G. Massey, PhD, 1995].

Decorated borders

The more amusing point is to understand what happens beyond the distortion threshold [13,14]. Here, there is a very remarkable effect. The more you pull, the more the willow stretches horizontally: as shown in figure 21, it is confined to a region of diameter D that becomes thinner and thinner. The force and the diameter are always related by a relation which was first derived by P. Pincus:

$$fD = kT$$

Ultimately, the stretching stops and a special regime is generated. When the diameter D becomes as small as the diameter of one tube, D^*, there are no knots left, and we lose most of the friction. Then immediately the willow is less tormented, and so it begins to spring back. As it springs back it opens more, and the friction is restored. Thus there is a so-called *marginal state* to which the system adjusts, and where it remains, with $D \sim D^*$. This marginal state has a constant force associated with it. The force per chain in this state, f^*, is related to kT and to the diameter $D \sim D^*$ through:

$$f^* = \frac{kT}{D^*} \tag{21}$$

A constant force per chain means a constant stress per unit area:

$$\sigma^* = vf^* \qquad (22)$$

where v is the number density of chains. So, what we are describing here is a system that, above a certain critical velocity, evolves to produce a constant shear stress. It is a very non-linear system — there is no longer a linear relation between velocity and stress — but if instead of the stress you consider the extrapolation length, you find that the extrapolation length should increase linearly with velocity:

$$b = \eta \frac{V}{\sigma^*} \qquad (23)$$

or in other words, according to a power law with exponent equal to unity. You may remember from the graph in figure 17 that the experimental exponent was originally something like 0.86. Actually, Léger *et al.* found that if they incorporate shear thinning in the viscosity η, they return to an exponent unity in their plot $b(V)$ — as expected from equation (23).

The relation between stress and velocity is shown in the graph in figure 22. Essentially, the stress increases with velocity and then saturates at the critical value V^*. Of course, if you go to very high velocities, then even with the straight tube without any entanglement there is some friction, and this corresponds to a higher branch above V_2, though this branch is proba-

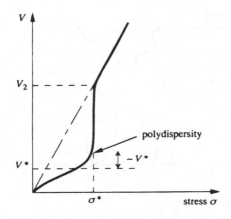

Figure 22. Shear stress versus velocity for
a polymer chain.

bly unobservable. The main practical feature is the
long straight line of constant stress in between V^* and
V_2.

One reason you might be worried about this analysis is that the length of the chains is not fixed — there
is a distribution of chain lengths, as is clear from figure 23. Indeed, this has the result that the sharp turn in
the curve at the critical velocity is broadened. Nevertheless, in practice V^* is so small compared with V_2
that we keep a very large plateau.

Will this plateau be important for practical purposes? Frankly, we don't know yet. There are all sorts

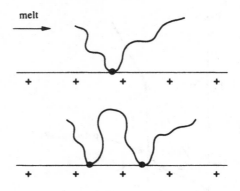

Figure 23. Distribution of chain lengths.

of remarkable instabilities which occur in dye flows of polymers, some of which generate very poor materials – for instance, the 'sharkskin' instability, which really makes a mess out of a plastic sheet or plastic tube. We don't know yet whether the slippage transition will be relevant or not. But I am pretty sure that the tool – anemometry near walls – will be very useful for the future.

French students are inclined to think that theory is the thing, and experiments are some sort of lousy cookbook operation which you do hastily at the end of your work. But look at the numbers. For the wind in the willows, theory represented maybe three months' work for four or five people; the experiments represented something like four years of work on the optical side, and four years of work on the surface side, done by different people. So experiments are really the stumbling block.

IV
Principles of adhesion

The wind in the willows was an example of soft interfaces in slippage. I would like to proceed now to a different situation, where the two partners are *separated* — that is, the problem of adhesion (figure 24).

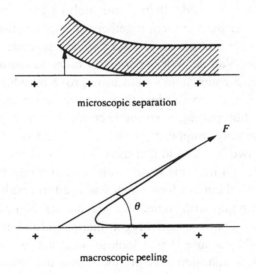

microscopic separation

macroscopic peeling

Figure 24. Adhesion of two separated interfaces.

Adhesion is an old science (table 1). I am particularly impressed by the Phoenicians who first invented a material that was both a glue and a seal. This apparently was the basis of the Phoenician navy's success.

Although adhesion has a brillant past, to a newcomer it may appear to be cookbook science. It is hard to teach a course on adhesion because it is a mixture: inventive chemistry, mechanics, physics, and other things as well. The unity probably lies in polymer science, because all adhesives are made of polymers.

Also striking are the orders of magnitude involved. If I take a typically non-adhering system, something like my two hands, then (fortunately) I can separate them after putting them together. This means that the energy per unit area G involved when I separate them is a pure Van der Waals energy of order 50 mJ/m^2. If you are a chemist and want to improve on adhesion, you have the naive view that you will automatically succeed by putting in strong chemical bonds. Now do the counting. Suppose you have chemical bonds between two blocks and that there is one bond for every $(3 \text{ Å})^2$, and that this bond is one electron-volt in energy. Working out how much you need to break that, you come up with something which is of the order of 1 J/m^2. That's much bigger than before, but still smallish, because if you look at what the experts in cookbook adhesion have produced over the years, you find that nowadays they easily produce 1000 J/m^2.

Principles of adhesion

Table 1. Glues through the ages.

~4500BC	sickles	(Judea)
		wood/tar/silex
~1300BC	*molten adhesives*	
	arrow/sulfur/arrowhead	(Egypt)
	adhesive sealants	
	wood/turpentine	(Phoenicians)
	wax	
	tar	
	cloth/latex	(Olmecs)
AD1040	printing characters	(China)
	wood/resin/clay	
~1900	nitrocelluloses→ shoes	
	rubber and solvent	
1912	bakelites	wood
	(formol-phenol	
	↓ urea-phenol)	
1940	*polyurethanes→*	
	metal/metal	
	epoxy	
	silicones	→ metal/glass
1950	*'hot melts'*	→ books
		→ shoes
1960	*polyimides*	→ high
		temperatures

45

1 The electrostatic model of adhesion

What is the source of these strong energies? The Russian school of Derjagin proposed one idea, based on observations using dissymmetric contacts. For instance, they would peel off a film of cellulose nitrate from a silicon wafer. This gives adhesion energies $G \sim 100$ J/m². They observed that inside the gap between the peeled film and the solid, sparks would show up. They concluded that electrostatic effects are important. If the redox potentials of the two sides are different, some electrons may be transferred at the moment of separation. The net result is then one sheet with a charge density σ_e, the other sheet carrying the opposite density $-\sigma_e$. In between, we have a large electric field:

$$E = 4\pi\sigma_e \tag{24}$$

and thus a large attractive force F between the two sides:

$$F = 2\pi\sigma_e^2 \tag{25}$$

If this process persists up to a certain final thickness of the air gap ($h_{\max}$), it leads to a separation energy:

$$G = Fh_{\max} \tag{26}$$

which can become large. The crux of the matter is to find $h_{\max}$.

Derjagin then returned to the classic literature on electron avalanches in gases under electric fields. If one electron is free and accelerated by the field, after a certain mean free path l it hits a gas molecule and creates more electrons (this mean free path is itself inversely proportional to the gas pressure). Avalanches show up when l is smaller than the gap thickness h. But if $h \ll l$, they are largely suppressed. Thus one arrives at $h_{max} \sim l$. In air under atmospheric pressure this leads to values of h_{max} which are large on the atomic scale: a few thousand angstroms. The conclusion is that this process can lead to large values of G, at least if the system is strongly dissymmetric (large σ_e).

This Russian model has not been well accepted in the rest of the world: (*a*) there is no systematic difference between symmetric junctions (where $\sigma \equiv 0$ by symmetry) and asymmetric junctions; (*b*) the model predicts huge G values at low gas pressures, and this is not really observed.

Where is the flaw? I suspect the crucial (hidden) assumption was that the peeled surface remained planar. We all know that Scotch tape, just after separation from a solid, displays fibrils. This probably occurs also at microscales: the final surface is very rough, and we may expect that the charges migrate (by hopping processes) to the tips of the profile. Ultimately, there is a large concentration of electric fields near

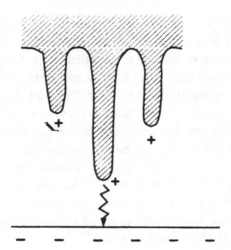

Figure 25. Electrical discharge from a fibril
may be the cause of the difference between
experimental observations and the
predictions of the Russian model.

these tips, and a discharge can start from these points, at values of the gap h which are still very small (figure 25). So I tend to think that the electrostatic contribution has been overestimated by the Russian school.

But this should not stop us. I am convinced that electrostatic effects can play a leading role, if we plan for careful experiments with two main requirements:

(*a*) select dissymmetric systems with strong charge transfer abilities, i.e. with suitable electron donors and acceptors grafted on each side;

(*b*) prevent roughness on the fracture profile, by choosing adhesives which are not too deformable.

With these provisos, I am willing to bet that electrostatic effects can, and will, be displayed unambiguously in the future. But let me now return to more classical sources of adhesion.

2 Fracture energy of glassy polymers

Dissipation at the fracture tip is the dominant source of adhesion energy. Let me make this more precise by choosing one particular example: namely glassy polymers which *craze* under tension. (Typical examples are polystyrene and polymethyl methacrylate.) These materials are known to be relatively tough, that is they have large fracture energies G. But the detailed mechanism has been understood only recently – by a former Cambridge student, Hugh Brown.

When we bend a slab of polystyrene, it becomes white, because of crazes: this process has been studied intensively over the past twenty years, especially by Kambour and Kramer. The microstructure of a craze is represented as shown on figure 26. Fibrils (diameter $D \sim 300$ Å) are pulled out from the glassy matrix. Near the end of each fibril, we have a high tension, comparable to the yield stress σ_y; thus the polymer is locally fluidised and flows into the fibril. The length h of the fibril can reach very high values (microns).

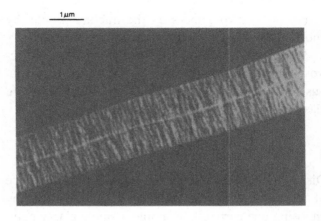

Figure 26. Craze in a random copolymer methyl methacrylate-
glutarimide.

Ultimately, if we pull hard enough, the fibril will break at some critical length h_c. Following Brown, we can find a scaling law for h_c by a simple argument:

(i) Far from the fracture tip, the stress stabilises to a constant value $\sim \sigma_y$ (required to pull out the fibrils).

(ii) Near the tip we have a stress concentration, ruled by the standard law for elastic media [15]: at a distance x from the tip, the stress diverges like $x^{-1/2}$.

(iii) This divergence is cut off at the first fibril ($x \to 0$) by the minimal x available ($x \sim D$).

(iv) At this last fibril, the stress reaches the critical value for rupture $\sigma_c \sim U_b / a^3$ where U_b is a bond energy, U_b/a a chemical force for rupturing one bond

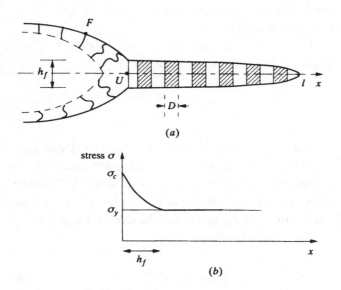

(a)

(b)

Figure 27. Stress on fibrils at a fracture.

(~1 nanonewton) and a^2 the average area per bond. Thus the stress scales like:

$$\sigma(x) = \sigma_c \left(\frac{D}{x}\right)^{1/2} \tag{27}$$

(v) The stress concentration, for fibrils of length h_c extends only up to distances $x \sim h_c$. This expresses a general feature of elastic fields, ruled by Laplacian equations, where modulations of wavelength λ along the horizontal axis penetrate only up to lengths $\sim\lambda$ in the vertical direction.

Thus, at $x \sim h_c$, we must cross-over to the yield stress σ_y. This imposes the condition:

$$\sigma_y = \sigma_c \left(\frac{D}{h_c} \right)^{1/2} \qquad (28)$$

This allows us to determine h_c, and finally to estimate the fracture energy G. One given fibril is extended (from its birth up to its final length h_c) under a stress σ_y (except for the minor region of stress concentration near the tip). Thus the work done in pulling fibrils is (per unit area):

$$G \approx \sigma_y h_c = \left(\frac{\sigma_c^2}{\sigma_y} \right) D \qquad (29)$$

The ratio σ_c / σ_y is roughly the ratio of chemical energies to Van der Waals energies (of order 50). But the presence of σ_c^2 in the formula for G is striking: it makes G very high. Equation (29) is the key to understanding both the toughness of simple plastics like polystyrene, and the adhesive performances of many glassy polymers. It is a good example of irreversible work performed near the fracture tip.

There are other examples. We shall describe one of them now [16], in connection with rubbers.

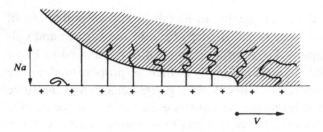

Figure 28. Mixing of grafted chains from a positive solid mixing with facing rubber.

3 Dissipation enhanced by flexible connectors

A typical situation is shown on figure 20, where we have a rubber facing a passive solid (e.g. glass). If we do nothing more, the adhesion energy is frightfully weak – dominated by Van der Waals forces. The trick is then to graft some chains to the solid, and ensure that they do penetrate into the rubber, generating what I call 'mushrooms'. Of course, we must have grafted chains which are chemically identical to the rubber, so that they will indeed mix. The resulting situation is represented in figure 28, where the rubber is shown on top, the glass at the bottom, and a fracture is advancing through the interface between them. We would like to know the energy of this fracture, at least at low speeds.

The mushrooms shown are initially at rest. As the fracture advances, each mushroom is stretched until it reaches its maximum length; it is then pulled out of the rubber, and it snaps back to the surface and collapses. To describe this situation, you need to know the maximum length (which is proportional to the number of units, N, in a grafted chain) and the force that is required to convince the chain to leave its nice rubber environment and to go into thin air. To determine the latter, one notes that there are two sources of unhappiness for a chain: one is Van der Waals energy (when it is pulled out, it has lost all its Van der Waals attraction to many neighbours); and the other is entropy (it has deformed and it has lost entropy). Now, in everyday life, Van der Waals energies and kT are about the same size, and their contributions to the force are comparable, so I will ignore the difference. The scaling form of the force is then simply a Van der Waals energy U divided by a monomer size a:

$$f^* = \frac{U}{a} \tag{30}$$

This is the threshold force. Once you know it, you can compute the energy involved in the fracture process. One of the few things which we have learnt at school is that work is a product of force by displacement. In our case, the force, if you are just at threshold (at low velocities), is f^*, and the displacement is Na. This gives you the energy G that you have to spend. To get it per unit area, you multiply by the number v of connecting units:

$$G(V \to 0) = f^* Nav = vNU \qquad (31)$$

You can see in equation (31) that the separation energy is proportional to a Van der Waals factor and to the length of the connecting chains; that is why you gain a big factor. On the other hand, it is also proportional in this dilute regime to the number of connectors, and if you don't have very many connectors you do not gain much.

The description I have just given, which is mainly due to Elie Raphaël, applies provided you have few connectors, so that they all go nicely into the rubber and don't create problems by crowding. But clearly, you are tempted to go further, because the industrial interest in this problem is not to check a nice formula for a weak v, but rather to find the optimal value of v that will provide you with good adhesion.

So, what is the optimal value? For something like two or three years, we had a completely naive idea about this: namely that the best thing to do was to match the distance between two connector chains and the mesh size of the network. However, this is not quite correct, and I hope to show you why.

The case which turned out to be the simplest to discuss is that of a *rubber/rubber contact* with connector

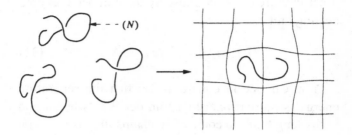

Figure 29. Finding the maximum concentration of free chains possible in rubber.

chains which are free to move and to bridge the interface (figure 9).

The first problem here is to find the maximum concentration of free chains which I can incorporate into my rubber. The starting point is the following experiment, depicted in figure 29.

I start with a melt of chains and I expose it to a rubber of the same chemistry (they are the same, so that there is no danger of segregation). I ask whether the chains migrate into the rubber or not, and to what extent. There were long disputes about this, and as is usual in such disputes, the conclusion was that the problem was ill-posed. It is not enough to say that you have a certain network here and now: you have to specify the history of the network with great care. If a network was cross-linked, for instance, in the presence of solvent, it has a certain natural relaxed state, or reference state, which is a swollen state. If I dry it up and later expose it to a reservoir of free chains, the chains will enter very naturally because this will allow the rubber to reconstruct its relaxed state. On the other hand, if the rubber was built by cross-linking from a dry state, i.e. with no solvent, then it is much more difficult for the chains to get in. I will focus on this latter case.

What you have, if you force chains in, is an elastic energy, because the network must swell. This elastic energy was first computed by Paul Flory. There are disputes about the exact numbers, but the structure of the theory is simple. The elastic modulus, E, is proportional to kT (because it describes an entropic elasticity) divided by the number N_0 of monomer units per mesh length, and also by a^3, the volume of one monomer unit:

$$E \sim \frac{kT}{N_0 a^3} \qquad (32)$$

Thus, the longer N_0, the weaker the mesh, as you would expect. The elastic energy we are talking about in this dry case is just a ϕ^2 energy:

$$\hat{F} = \tfrac{1}{2} E \phi^2 \qquad (33)$$

where ϕ is the dilation produced by the chains that are forced in. Once you know this, you can calculate the energy each chain has to suffer when it enters, due to this dilation. In more sophisticated language, this is the shift in chemical potential of the chains:

$$\mu = \frac{\partial \hat{F}}{\partial \left(\dfrac{\phi}{Na^3} \right)} = kT \frac{N}{N_0} \phi \qquad (34)$$

This is the energy you have to overcome: if it is larger than kT, the chain will never get in; if it is smaller than kT, the chain will be accepted. Thus, the maximum dilation at which the chains will be accepted is of order:

$$\phi_{max} = \frac{N_0}{N} \qquad (35)$$

I like equation (35) because it short-circuits a rather painful and detailed discussion on swelling of gels by polymers; the physics really is all here in this very simple formula (except for minor logarithmic factors).

Now, let me return to my rubber problem (figure 30). I know the adhesion energy when I know how many chains I have in the skin region. This in turn is simple counting, once I know the maximum concentration $a^{-3}\phi_{max}$ which is allowed in the bulk. So I multiply this maximum concentration by the thickness

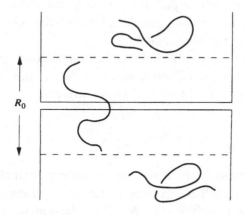

Figure 30. Calculating density of connecting
chains in the skin region.

of the skin region $R_0 = N^{1/2}a$ (i.e. a coil size, proportional to the square root of N), and when I do this, I
find the maximum density of connecting chains v_m
that I can achieve:

$$v_m = \frac{\phi_{\max}}{Na^2} R_0 \qquad (36)$$

or changing to dimensionless units, va^2:

$$v_m a^2 = \frac{N_0}{N^{3/2}} \qquad (37)$$

v_m is proportional to the distance between cross-links,
N_0, and it goes like the inverse of a power of the mo-

lecular weight. Whenever we are below this grafting density, the chains can indeed build up the connecting links we are talking about.

This tells us what is the maximum adhesion energy we can hope to collect from the mobile chains: inserting $v = v_m$ into equation (31), we arrive at:

$$G_{max} = \frac{U}{a^2} \frac{N_0}{N^{1/2}} \tag{38}$$

Here, surprisingly, it is better to use rather small connectors (small N). However, the whole discussion is meaningful only for $N > N_0$. Thus, the utmost value of G_{max} which we can hope to get in this case is of order $Ua^{-2}N_0^{1/2}$. Typically $N_0^{1/2} \sim 10$ and G_{max} is not very much enhanced over its Van der Waals value.

A couple of remarks should be made at this point:

First, the interaction energy due to elastic deformations in the network is given by equations (33) and (34). It is inversely proportional to N_0. The structure of this energy is very similar to what we have when we dissolve the N chains not in a rubber, but in a melt of shorter chains – the shorter chains being in fact of length N_0. As first shown by Sam Edwards, the short chains *screen out* the repulsive interactions between the long chains, and this reduces the effective excluded volume interaction by a factor $1 / N_0$. The similarity between a rubber matrix and a melt matrix is very striking. In the melt case, long-range screening proceeds by translation motions of the short chains, and the translational entropy per monomer is $\sim 1 / N_0$. In the rubber case, screening proceeds by deformation of network segments, and this gives an entropy which is also proportional to $1 / N_0$ (equation 32). Rubbers are screeners.

Secondly, our values of v_m and $G_{\max}$ hold only when the rubber was cross-linked in the dry state. But, if we want to have more free chains inside a network, there is a natural route: cross-link the network *in the presence of the mobile chains*, which then fit in naturally. This is indeed what happens with rubber bands for tyres, which contain a significant fraction of free chains, and stick together rather naturally.

Let us now return to the grafted chain problem of figure 28. When the grafting density v is small ($va^2N < 1$), the mushrooms do not overlap and we can add up their contribution (equation 31) safely. What happens at higher v? At some moment, the grafted chains will repel each other, and elongate normal to the surface, creating a 'brush' of thickness L.

The standard approach to these brushes [17] is based on a mean field calculation of the Flory type: the free energy F per grafted chain is of the form (ignoring coefficients):

$$\frac{F}{kT} = \frac{L^2}{Na^2} + \frac{N}{N_0}\phi \qquad (39)$$

where the first term is an elastic energy, while the second term is the chain–chain repulsion screened out by the rubber. Here, the volume fraction ϕ of grafted chains inside the brush is:

$$\phi = \frac{vNa^3}{L} \qquad (40)$$

Equation (39) holds if $L \geq N^{1/2}a$, i.e. when there is a significant stretching. Then, optimising F with respect to L, we get:

$$L = (va^2)^{1/3} N_0^{-1/3} Na \qquad (41)$$

Thus, stretching begins when:

$$v \geq v_s = a^{-2} \frac{N_0}{N^{3/2}} \qquad (42)$$

Note that the corresponding value of ϕ is just equal to the limiting ϕ discussed in equation (35) for mobile chains. It is also interesting to observe that there is an interval of grafting densities:

$$\frac{1}{Na^2} < v < v_s \qquad (43)$$

where the mushrooms overlap, but do not stretch because of the Edwards screening.

Increasing v beyond $v = v_s$, we do get a brush, with L increasing and ϕ increasing. Ultimately, at some moment, we reach an important limiting point: ϕ is of order unity. This, in the usual jargon, is called the *dry brush* limit. Returning to equation (40) for ϕ, we see that it occurs when $v = v_{max}$, where $v_{max}a^2 = N_0^{1/2}$.

Beyond this point, the brush decouples from the rubber matrix, and we lose most of the adhesion. This tells us the maximum adhesion energy we can hope to get from a brush: inserting $v = v_{max}$ into equation (31), we arrive at:

$$G_{max} = \frac{U}{a^2} \frac{N}{N_0^{1/2}} \tag{44}$$

So, we can expect a significant enhancement of adhesion if $N \gg N_0^{-1/2}$; a rather mild requirement.

At this point I must make three remarks:

First, the dynamics of interdigitation is very slow for tethered chains exposed to a network. Thus, the adhesive situation which we discussed occurs only if the rubber and the grafted surface have been incubated together for a long time.

Secondly, our discussion was restricted to 'unbound' systems: for instance, in the case of figure 30, the connectors are not chemically attached to the rubber. The opposite situation is also of interest: we can arrange that the terminal group of the connector carries a vinyl group, which can be permanently linked to some double bond in the rubber.

The adhesion energy for the 'bound' case is very similar to equation (31): all we need to do is to replace the Van der Waals energy U by a chemical bond energy U_b in equation (30) for the force or equation (31) for the energy. In practice, $U_b / U \sim 40$, and this represents a significant increase in adhesion. This gain by bonding was verified experimentally by Gent and co-workers on very short grafts. It is very much to be hoped that it will be tested for larger grafts in the near future.

Thirdly, equation (31) holds only when the fracture takes place at the interface, rather than in the form of cohesive fracture inside bulk rubber. The parameters which decide upon this choice are quite complex. But a first, naive hint is obtained by comparing the corresponding fracture energies.

The bulk fracture energy can be obtained by a formula similar to equation (31), where we look at the strands which must be cut by one dividing plane in the bulk. The strands have N_0 monomers. They involve a thickness $\delta \sim N_0^{1/2} a$ around the cutting plane. The number of monomers per unit area in this region is $a^{-3}\delta = N_0^{1/2} a^{-2}$. The number of strands ν_{bulk} is N_0 times smaller:

$$v_{\text{bulk}} = N_0^{-1/2} a^{-2} \qquad (45)$$

Thus:

$$G_{\text{bulk}} = v_{\text{bulk}} U_b N_0 = \frac{U_b}{a^2} N_0^{1/2} \qquad (46)$$

With free connectors, we find that $G < G_{\text{bulk}}$ in most cases. Thus we expect that equation (31) holds systematically. But with bound connectors, G becomes higher than G_{bulk} beyond a rather low grafting density:

$$v_{\text{lim}} = N_0^{1/2} N^{-1} a^{-2} \qquad (47)$$

It is often true (although not rigorously required) that fracture takes place by the scenario associated with the lowest G. If we accept this prescription for adhesion purposes, with bound connectors, we see that there is no interest in grafting densities higher than v_{lim}.

4 Far field contributions to the adhesive energy

Here we shall be concerned with adhesives which are *poorly cross-linked rubbers*. This means that inside the network we have many chains which are free; and many chains which are tied at one end only. In cases

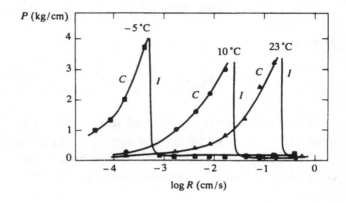

Figure 31. Peel force *P* versus peel rate *R* for an un-cross-linked butadiene-styrene rubber adhering to a PET polyester film. The symbols *C* and *I* denote cohesive failure and interfacial failure, respectively. (A. Gent, R. Petrich, PRS London **A310** 433, 1969)

like this, the low frequency modulus μ_0 (related to the network) is small. But the high frequency modulus μ_∞ (which contains the effects of the entangled free chains and of the dangling ends) is high. Typically, we can achieve:

$$\lambda = \frac{\mu_\infty}{\mu_0} \sim 100 \qquad (48)$$

It was shown by Gent and Petrich that poorly cross-linked systems of this type have a very anomalous curve $G(V)$ (adhesive energy / velocity) as shown on figure 31. The anomaly disappears upon further 'curing' (when the network is more cross-linked, and μ_0 raises to become comparable to μ_∞).

Here we shall try to give a simple interpretation of this effect. The major idealisation which we do is to assume that mechanical relaxation inside the network is described by a *single* relaxation time τ. Of course, this is very crude: the dangling ends have a wide distribution of length, etc. But the one time approximation allows to understand what happens rather easily.

The simplest formula for the complex modulus $\mu(\omega)$ as a function of frequency (ω) is the following:

$$\mu(\omega) = \mu_0 + (\mu_\infty - \mu_0)\frac{i\omega\tau}{1 + i\omega\tau} \qquad (49)$$

Usually, with one relaxation time τ, we think that we have to cope with two regimes, $\omega\tau > 1$ and $\omega\tau < 1$. Here, in fact, we have *three* regimes:

(i) At very low ω, $\mu = \mu_0$, we expect a *soft solid*.

(ii) When $1 > \omega\tau > \lambda^{-1}$, we can approximate:

$$\mu(\omega) \sim (\mu_\infty - \mu_0)i\omega\tau = i\omega\eta \qquad (50)$$

The modulus is purally imaginary: we are dealing with a *liquid* of viscosity:

$$\eta = (\mu_\infty - \mu_0)\tau \sim \mu_\infty\tau \qquad (51)$$

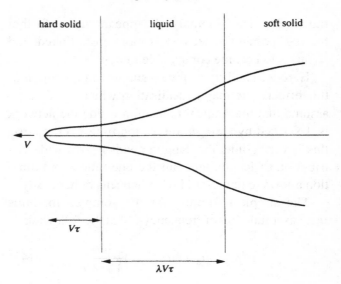

Figure 32. A fracture profile spanning three regions:
hard solid, liquid, soft solid.

(iii) At high frequencies ($\omega\tau > 1$) we recover a *strong solid* ($\mu = \mu_\infty$).

What are the consequences of this on a moving fracture? When the fracture velocity V is not too small, so that $V\tau$ is larger than the size of the fracture tip, we can think in terms of a continuum, and we find three regions in our rubber (figure 32):

– Near the fracture tip, we are discussing small spatial scales, or short times, and we have a strong solid.

— At intermediate distances, $V\tau < r < \lambda V\tau$, we have a liquid, giving a large dissipation.

— At higher distances $r > \lambda V\tau$, we have a soft solid.

You might think that the effects in the liquid zone are not very important, because the stresses here are relatively small — we are far from the fracture tip. On the other hand, because r is large, we have a huge volume of fluid-like behaviour and the overall dissipation is increased. It will turn out that this volume factor dominates.

Although we are talking about a complex visco-elastic medium, the scaling law for the stress as a function of distance is still simple, and equivalent to what we have in a simple elastic medium:

$$\sigma(r) \sim K_0 / r^{1/2} \qquad (52)$$

where the factor K_0 is associated with the adhesion energy G_0 due to local processes near the tip. Again ignoring all coefficients, we have the standard relation from fracture mechanics:

$$G_0 \cong \frac{K_0^2}{\mu_\infty} \qquad (53)$$

I shall make three remarks concerning the stress distribution:

First, equation (52) is simply a scaling law, ignoring all specific details for different components of the stress: for instance, on the fracture tip, the normal stress component σ_{zz} must vanish identically – but the other components still follow the scaling form (52).

Secondly, why does this simple form remain valid in a viscoelastic medium? The equations of motion, in terms of density ρ and local velocity v, reduce to:

$$\rho \frac{\mathrm{D}v}{\mathrm{D}t} = -\nabla\sigma \cong 0 \tag{54}$$

where we set $\mathrm{D}v/\mathrm{D}t = 0$, because we are dealing with velocities much smaller than the sound velocity: the inertial terms are negligible. Thus the equations are simply $\nabla\sigma = 0$, and are the same for any viscoelastic medium. The stress components must also satisfy compatibility conditions (because they derive from a displacement field). But these geometric conditions, again, are the same for any viscoelastic medium.

Thirdly, in an elastic medium of modulus μ, with a stress field described by equation (52), the displacement field u is simply related to the stress by the scaling law:

$$\sigma = \mu\nabla u \tag{55}$$

and this imposes:

$$u = \frac{K_0}{\mu} r^{1/2} \tag{56}$$

If we take the scaling structure of the product (σu) along the fracture (where u now measures the opening of the fracture), we find it is independent of distance:

$$\sigma u = \frac{K_0^2}{\mu} = G_0 \qquad (57)$$

where G_0 is the corresponding adhesion energy. For our more complex problem, we shall again find the separation energy G from the product (σu), calculated at larger r (in the soft medium): measuring at large distances we incorporate all dissipation effects.

Let us now return to figure 24 and find out the overall shape of the fracture profile $u(x)$. In the strong solid region, we have a classical parabolic shape:

$$u = \frac{K_0}{\mu_\infty} x^{1/2} \qquad (58)$$

and $\sigma u = G_0$.

In the liquid zone, σu is not constant. The scaling law relating u to σ is based on a viscous stress:

$$\sigma = \eta \frac{d}{dx}\left(\frac{du}{dt}\right) = \eta V \frac{d^2 u}{dx^2} \qquad (59)$$

and with $\sigma \sim x^{-1/2}$ (equation 52) this gives $u \sim x^{3/2}$. Thus the product σu increases linearly with x:

$$\sigma u = G_0 \frac{x}{V\tau} \qquad (60)$$

When we reach the soft solid region ($x = \lambda V \tau$), we find:

$$\sigma u = \lambda G_0 \qquad (61)$$

and this gives us the overall adhesion energy G:

$$\frac{G}{G_0} = \lambda = \frac{\mu_\infty}{\mu_0} \qquad (62)$$

A more rigorous derivation of equation (62) has been given by Hui.

The result deserves a number of comments. Note first that the viscoelastic corrections give a *multiplicative* effect: they do not add up a constant term to G_0. Secondly, we see that the ratio G / G_0 can be very large if the material is very poorly cross-linked ($\mu_0 \rightarrow 0$). This explains why the enhancement in G disappears upon further curing.

All this discussion holds for a thick glue (bulk fracture). But very often the glue is in the form of a thin slab (thickness W) as shown on figure 33. The opening process is then cut off at distances $x \sim W$. Thus when $\lambda V \tau > W$, the dissipation zone is restricted in size and we have:

$$G = \frac{G_0 W}{V \tau} \qquad (63)$$

72

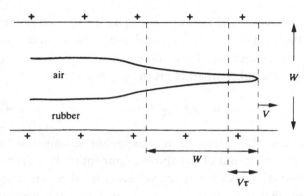

Figure 33. A thin slab of glue with thickness W.

In this regime, $G(V)$ is a decreasing function: the pulling force drops if the velocity increases. This often generates mechanical instabilities in the fracture process. We believe that this instability is the source of the peak in the experimental $G(V)$ curve observed by Gent and Petrich.

What happens when we have not a single relaxation time τ, but a broad distribution of relaxation times, giving a complex form to the dissipative modulus $\mu(\omega)$? A natural attempt, proposed long ago by many authors, amounts to assuming that $G(V)$ is proportional to the modulus $\mu''(\omega)$ measured at a frequency $\omega = V / l$, where l is some characteristic length of the rubber network. However, this is not correct in general, as pointed out, by Langer and Barber in particular, on simple examples. It does hold in one case:

73

namely when the network is *just at its gelation point*: at the transition between isolated clusters and one infinite cluster plus finite molecules. At this point, the elastic modulus $\mu(\omega)$ is given by a power law:

$$\mu'(\omega) = \mu''(\omega) \cong \omega^{\alpha} \qquad (64)$$

where α is a certain critical exponent. In this case, a simple extension of the above argument (with a distribution of relaxation times which is fixed by α) does show that $G \sim V^{\alpha}$. This appears to be an exceptional situation; however, this region near the connectivity threshold is that of long τ and large G: many practical materials may be in the region.

5 Extension to 'tack'

Certain polymer melts (un-cross-linked) are sticky. This is called *tack* and is an important feature for certain adhesion purposes. For instance, if we establish contact between a fluid elastomer and a metal, we find that to separate them we need energies G which are high – much higher than the Van der Waals energy.

It is tempting to relate this remarkable effect to the viscoelastic properties of the elastomer. The proposed mode of separation is shown on figure 34. A fracture propagates along the solid, with velocity $V \sim L\theta^{-1}$, where L is the contact size and θ the separation time.

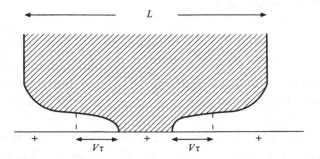

Figure 34. A fracture propagates along the solid, with velocity $V \sim L\theta^{-1}$, where L is the contact size and θ the separation time.

We again have a liquid zone, ranging from $x = x_1 = V\tau$ up to L. The adhesion energy is:

$$G = \sigma u|_{x=L} = G_0 \frac{L}{x_1} = G_0 \frac{\theta}{\tau} \qquad (65)$$

Thus if $\theta \gg \tau$, we expect a large enhancement. A number of comments should be made here:

First, another presentation of the result (equation 65) is:

$$G = G_0 \frac{\mu_\infty}{|\mu(\omega)|} \qquad (66)$$

where $\mu(\omega)$ is the analog of equation (49) for a visco-

Principles of adhesion

elastic liquid ($\mu_0 = 0$):

$$\mu(\omega) = \frac{i\eta\omega}{1 + i\omega\tau} \approx i\eta\omega \quad (\omega\tau << 1) \qquad (67)$$

and $\omega = \theta^{-1}$ is the separation frequency. In the form (66) this tack property is a natural generalisation of equation (62).

Secondly, it is important to observe that the enhancement factor θ / τ is large if the elastomer has a rather low molecular weight (low τ). Of course, there is a limitation to this: if we choose chains which are so short that they do not entangle, μ_∞ drops down, and we lose the enhancement factor μ_∞ / μ_0 in equation (62).

Thirdly, our material is liquid: at very large times a liquid collapses under its own weight. The collapse time t_c may be estimated as follows. The shear rates are of order $1 / t_c$ and the shear stresses are η / t_c. They must balance the gravitational pressure, which for a sample of size L is of order $\rho g L$ (ρ = density, g = gravitational acceleration). This gives a maximum value for the enhancement factor:

$$\left.\frac{G}{G_0}\right|_{max} = \frac{t_c}{t} \approx \frac{\mu_\infty}{\rho g L} \approx 10^3 \qquad (68)$$

It may be, however, that the accelerations involved in

76

Figure 35. A contact zone breaking up into a bundle of elastomer fibrils.

separation process are $g_{eff} \gg g$, and that g_{eff} should be used in this estimate. Note on equation (68) that if μ_∞ becomes small (with chains smaller than the entanglement length) the enhancement disappears.

Fourthly, a nice feature of equation (65) is that it is independent of the size L of the sample: thus the result may hold even if the contact zone breaks up into a bundle of elastomer fibrils (figure 35).

On the whole, we have an attractive (but tentative) picture of tack based on viscoelastic effects: this should be compared to careful experiments at variable θ.

The main practical use of tack is in *rapid adhesion*: when we must bind objects very fast (e.g. assembling books on a production line). The materials used are commonly known as 'pressure-sensitive adhesives'. They are based on (nearly) un-cross-linked chains (plus some additives). The adhesive joint is prepared by applying the glue under a certain weak pressure. What is the role of this pressure? It improves the contact on a rough solid surface, as shown in figure 36.

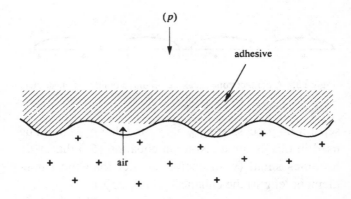

Figure 36. Pressure improves contact on a rough surface.

There are some 'hollow' regions where contact is not initially realised, and must be encouraged by the pressure p. Early descriptions of this filling process were based on a purely viscous model, where the adhesive flows towards the hollow regions like a liquid. Recently (1995), C. Creton and L. Leibler have pointed out that the viscoelastic features are essential, and that the length of time required for good contact can be better understood on that basis.

V
Polymer/polymer welding

We have seen that the toughness of bulk (glassy) polymers, which craze under tension, begins to be understood through an original idea of H. Brown [18]. We shall now try to extend the Brown ideas to various systems of 'weak junctions'. The junction may be a partly healed contact between two identical polymer blocks A/A, as in the experiments of the Lausanne group [19,20]. Alternatively, it could be a contact between two different polymers A and B.

In all our discussions, we shall assume these junctions to be perfect, with full contact between the two partners, and no gaps. Experimental arguments for the existence of these good contacts have been presented by Kausch and coworkers [19].

Our aim here is:

(*a*) to give a brief reminder of the theoretical description of the weak junctions;

(*b*) to show how some basic mechanical properties can be related to the structure.

One of the major conclusions, for the A/A case, is

that *chain ends* play a crucial role. Thus, any attraction between a chain end and the free surface of one A block will react significantly on the A/A mechanical properties after welding.

This type of attraction was first suggested by systematic experiments on melts by D. Legrand and G. Gaines [21], showing that the surface tension γ of oligomers was often lower than the surface tension γ_∞ of a high polymer, and that the correction has the form:

$$\gamma_x - \gamma(N) \sim N^{-x} \tag{69}$$

where N is the degree of polymerisation, and x an exponent of order 2/3. The fact that $x < 1$ shows that we are not dealing with a simple uniform dilution of chains ends (which would give a correction $\sim N^{-1}$). The most natural way of understanding the Legrand–Gaines result amounts to assuming that the chain ends are attracted to the surface. The IBM group [22] has argued that a typical monomer along the chain suffers an entropy loss of order unity when it is located near the free surface, because the chain is 'reflected' here, while the chain ends do not have this loss: thus one expects, on purely entropic grounds, a gain of free energy $\sim kT$ for each chain end brought to the surface. There are also enthalpic effects which may increase or decrease the surface attraction. But if on the whole the attraction is of order kT per chain end, we reach a simple regime [23], where all chains within one radius of gyration $R_0 = N^{1/2}a$ of the surface put their ends on the surface, and the deeper chains are unperturbed. This leads to a surface fraction of chain ends ϕ_s of order:

$$\frac{2}{N} \cdot \frac{R_0}{a}$$

(where a is a monomer size), and thus:

$$\phi_s = \sim N^{-1/2}$$

(or $x = 1/2$) in this regime. We shall call this the *normal attractive regime*. (The value $x = 2/3$, observed by Legrand and Gaines, may be the result of a cross-over between zero attraction and normal attraction.)

This interpretation of the Legrand–Gaines results is still controversial. For instance, Dee and Sauer interpret $\gamma(N)$ not as an effect of chain ends, but from the empirical N dependence of the (P,V,T) equation of state for oligomers (the main feature here being the change of the equilibrium density $\rho(N)$). They use a standard mean field analysis of the interfacial energy, as related to the equation of state and to the range of the intermolecular forces (the magnitude and the range being assumed independent of N). Dee and Sauer get remarkable fits to the Legrand–Gaines data, without involving any special localisation of chain ends! Here, however, we shall keep in mind constantly the possibility of chain-end segregation near the free surface: indeed, we shall see that some of the neutron data on partial healing of A/A interfaces are more easily understood in the normal attractive regime than in zero attraction.

1 Healing of an A/A interface

The basic healing experiment is idealised in figure 37. We start with two blocks of the same polymer, which we call H and D. [For certain experiments, D may be a deuterated polymer, while H is the usual proton-carrying species.] The two blocks are put into close contact under a mild pressure, at a temperature close to the glass point T_g, during a time t. The polymer chains from H and D begin to intertwine, and build up a diffuse profile for the D concentration ϕ_D (figure 38). We are interested here primarily in this interdigitation process, at times t smaller than the reptation time of the chains T_{rep}. This corresponds to spatial widths of the profile $e(t)$ which are smaller than the coil radius R_0.

Most experiments have been performed with polystyrene, and with H and D chains of comparable length: $N_H \cong N_D = N$. The choice of N is non-trivial:

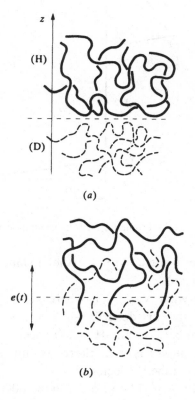

(a)

(b)

Figure 37. An idealisation of the basic healing experiment.

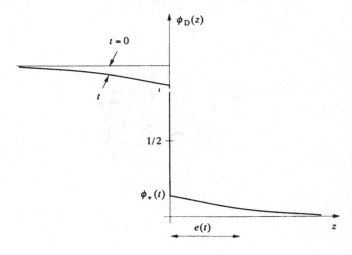

Figure 38. Profile for the D concentration ϕ_D.

(*a*) We want $N \gg N_e$ (the distance between entanglements).

(b) We want $N\chi_{HD} < 1$, where χ_{HD} is the (small) Flory parameter describing a weak trend for segregation between the H and D species. When this condition is satisfied, there is no significant segregation of the D species.

Typically N will be of order 2000–6000, while N_e ~ 300. The thickness $e(t)$ of the partly healed zone is in the range of 100 Å – too small to be studied by forward scattering of charged particles. The main experimental tools used to measure the healing profile have been SIMS and neutron reflectance.

Most data do show that the overall thickness $e(t)$ grows like $t^{1/4}$:

$$e(t) \sim R_0 \left(\frac{t}{T_{\text{rep}}} \right)^{1/4} \qquad (t < T_{\text{rep}})$$

This is the natural law for spatial motions of one labelled monomer in an entangled melt [24]. After a time t, the chain carrying this monomer has moved along its own tube by a curvilinear length:

$$s(t) = (D_{\text{tube}}t)^{1/2}$$

where D_{tube} ($\sim N^{-1}$) is the tube diffusion coefficient. The corresponding distance as the crow flies is:

$$e(t) = (d\,s(t))^{1/2} \qquad (s > d) \qquad (70)$$

where $d = N_e^{1/2}a$ is the tube diameter, and (70) co-incides with the empirical law for $e(t)$.

However, this simple agreement ignores the important fact that near the contact surface the chains were originally reflected, and thus most of the tube motions do *not* give any intertwining. A theoretical reflection on the problem was performed long ago by various authors [25,26,27], and will be summarised here.

85

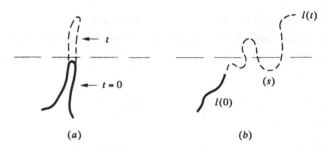

Figure 39. 'Hairpin' processes in adhesion.

(*a*) For $N \gg N_e$, it is reasonable to assume that the 'hairpin' processes of figure 39 are negligible. The entropy of a hairpin on a lattice model is one half of the entropy of a free chain more generally: hairpins are disfavoured by a factor of order $\exp(-n/2N_e)$, where n is the contour length of the hairpin.

(*b*) Then, at the time of interest (where $e(t) > d$), all the interdigitation is due to the motion of chain ends; one of them will start from some initial position (within $e(t)$ of the interface), and may cross the interface one or more times. The number of monomers which it brings to the other side is a fraction of $s(t)$. Therefore, the total number v of monomers D going through the interface (per unit area) is of the form:

$$v \cong s(t) \int_{-e}^{0} \phi_e(z) \, \mathrm{d}z \qquad (71)$$

where $\phi_e(z)$ is the initial distribution of chain ends. In

86

the original discussions [25,26,27], it was assumed that $\phi_e(z)$ is uniform $\phi_e(z) = 2/N$. But nowadays we know that chain ends may have been attracted to the original free surface of the block: as explained above, in normal attractive conditions, this will bring another (dominant) contribution to the integral, proportional to $\phi_{s0} = N^{-1/2}$. Thus we have two cases:

$$\left.\begin{array}{ll} v \sim s(t)e(t)N^{-1} & \text{(no attraction)} \\ v \sim s(t)aN^{-1/2} & \text{(normal attraction)} \end{array}\right\} \quad (72)$$

(*c*) Because of the reflection of chains at the original interface, the profile is *discontinuous* (for spatial intervals larger than the tube diameter d). The general aspect is shown on figure 38. Of major interest is the concentration $\phi_+(t)$ of D monomers, on the H side, for $z \to 0$. We may write:

$$v \sim \phi_+(t)e(t) \quad (73)$$

Comparing (72) and (73), using the normalisation factors, and inserting $\phi_{s0} \sim N^{-1/2}$, we then arrive at:

$$\left.\begin{array}{ll} \phi_+ \sim \left(\dfrac{t}{T_{\text{rep}}}\right)^{1/2} & \text{(no attraction)} \\[3ex] \phi_+ \sim \left(\dfrac{t}{T_{\text{rep}}}\right)^{1/4} & \text{(normal attraction)} \end{array}\right\} \quad (74)$$

87

Thus, when chain ends were originally numerous at the surface, $\phi_+(t)$ rises more rapidly.

On the experimental side, recent data on the profile comes from the neutron reflectance experiments of Reiter and Steiner [28]. They found that their profiles could not be described by simple diffusion (giving an error function); but they could be described by the superposition of *two* error functions E_{slow} and E_{fast}:

$$\phi_D(z) = 2\phi_+(t)E_{slow}\left(\frac{z}{e(t)}\right) + (1 - 2\phi_+)E_{fast}\left(\frac{z}{\sigma_c(t)}\right)$$

$$(75)$$

where the error functions $E(z)$ are normalised by $E(0) = 1/2$, $E(-\infty) = 1$, $E(+\infty) = 0$.

For the 'fast' component (describing what we called the discontinuity), they found $\sigma_c(t)$ very weakly dependent on time, increasing from ~20 Å to 30 Å in the time interval $0 < T_{rep}$. For the 'slow' component, the result is $e(t) \sim t^{0.17}$ – not too far from equation (70).

But their most interesting result is related to $\phi_+(t)$. They found $\phi_+(t) \sim t^{0.22}$, very close to the prediction of equation (74) for normal attraction between chain ends and the free surface. What is nice is that they obtained this without being biased by any theoretical prediction!

Thus the Reiter–Steiner experiment does suggest that (in their conditions of sample separation) chain ends were originally attracted to the surface.

Kausch and coworkers [19,20] have measured the fracture energy G of partly healed A/A contacts: after healing over a time t, the sample is brought back to room temperature, where it is glassy, and then fractured along the junction. Experimentally, in most cases, the fracture energy G increases with healing time: $G \sim t^{1/2}$ ($t < T_{\text{rep}}$).

We have seen earlier (equation 29) how the toughness of glassy polymers in bulk can be explained following the ideas of H. Brown. Let us now transpose this to partly healed interfaces.

Here the number of bridging chains per unit area is expected to be proportional to $\phi_+(t)$: any D monomer which has just crossed the border has a finite probability of being directly linked to the D side. Thus, to describe chemical rupture of the fibrils, we should perform the replacement:

$$\sigma_c \to 2\phi_+(t)\sigma_c \qquad (76)$$

where the factor (2) is fixed by the condition that σ_c returns to its bulk value at $t > T_{\text{rep}}$ ($\phi_+ \to 1/2$). Equation (29) then gives:

$$G(t) = G_{\text{bulk}} \cdot 4\phi_+^2(t) \qquad (77)$$

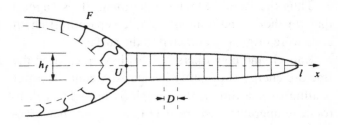

Figure 40. Separation of two blocks
by fibril rupture.

If, and only if, chain ends were originally numerous at the surface, we can return to equation (74) and write $\phi_+(t) \sim t^{1/4}$, giving the experimental form $G \sim t^{1/2}$. Thus the Kausch law, combined with the Brown model, does suggest that a large number of chain ends were available at the interface when healing started.

Some of you may be worried by the following point: in the Kausch experiments, the original blocks H and D were in fact obtained by rupture of one single sample. Could it be that chain ends were *very numerous* ($\phi_s > N^{-1/2}$, possibly $\phi_s \sim 1$) on the interface at $t = 0$? We do not believe this to be the case, as explained on figure 40.

The separation of the two blocks took place via fibril rupture, but after this, the half fibrils on both sides have probably retracted to build again a compact layer of polymer on each lip of the fracture (around point F): in this retraction process, chain ends may be buried in each layer. Most of the chains in this layer belonged to portions of the fibrils which were not disrupted chemically. Thus, if the retraction led to an equilibrium, we again expect $\phi_s \sim N^{-1/2}$.

2 A/B interfaces

The qualitative aspect of an A/B interface is shown on figure 41. A simple understanding of the structure can be obtained, starting from an abrupt interface, and allowing one A chain to protrude in the B side (figure 41). If m monomers are exposed in this process, the enthalpy required is:

$$\Delta H_m \sim m \chi kT \qquad (78)$$

where χ is the Flory parameter [21] describing AB mixtures. The average value of m corresponds to $\Delta H_m \sim kT$, and is thus:

$$(\overline{m}) = \chi^{-1} \qquad (\chi < 1) \qquad (79)$$

(We constantly assume that $\overline{m}$ is much smaller than the overall chain length N.)

Since the protruding chain is a random walk, the

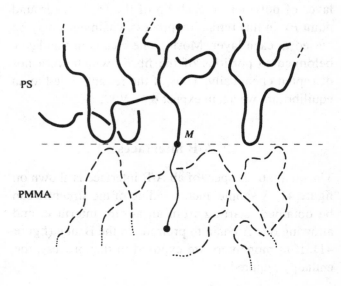

Figure 41. One chain on the A side is protruding
into the B side.

width *e* of the interface is the size of this random
walk:

$$e \cong a\overline{m}^{1/2} = a\chi^{-1/2} \tag{80}$$

and *e* is much larger than *a*, if χ is small: we shall
constantly focus on this limit. Of course, the result
(equation 80) can be derived by more rigorous means,
but the present approach is often illuminating.

The distribution of m values is the Boltzmann exponential:

$$p_m = \frac{1}{\overline{m}} \exp\left(\frac{-\Delta H_m}{kT}\right) = \frac{1}{\overline{m}} \exp\left(\frac{-m}{\overline{m}}\right) \qquad (81)$$

Of major interest for mechanical properties, is the probability that the protruding chain *entangles* with the surrounding matrix. If we define an average chemical distance between entanglements N_e, we may write for the probability f of entanglements:

$$f = \sum_{N_e}^{\infty} p_m = \exp(-N_e \chi) \qquad (82)$$

Of course, this formula is very approximate, because N_e need not be the same for the two partners A and B, nor for the mixtures: a certain weighted average would then be required. But equation (82) is still a reasonable starting point for discussing the mechanics of A/B contacts.

Long ago a remarkable series of experiments was performed by Iyengar and Erickson [29]. They measured the adhesion energy G of various polymers, on PET, by a 90° peeling test (at a fixed velocity of 5 cm/s). The results were plotted as a function of the Hildebrand solubility parameter δ. They show a dramatic drop of G as soon as the δ parameters differed by more than one unit.

Can we establish contact between these data and equation (81) for the probability of entanglements? Let us assume that:

(*a*) G is associated to the post-craze fracture of a glassy A/B junction;

(*b*) the entangled A or B chains in the junction must break.

Then we may apply Brown's equation (29), provided that the chemical rupture stress σ_c is suitably reduced: only the entangled chains at the junction contribute. This could give:

$$\sigma_c \rightarrow \sigma_c f$$

and:

$$G = G_0 f^2 = G_0 \exp(-2N_e \chi)$$

The result is an exponential drop in a $G_{1c}(\chi)$ plot: from the data, Iyengar and Erickson had proposed a different law: $G_{1c} \sim \exp\left[-k|\delta_A - \delta_B|\right]$. (Remember, for simple Van der Waals interactions, $\chi \sim (\delta_A - \delta_B)^2$.)

However, these differences are probably not very significant:

(*a*) As already explained, N_e need not be the same for all AB pairs: there need not be a universal plot $G_{1c}(\delta)$.

(*b*) PET is always partly crystallised, and this complicates the picture.

The essential point is the rapid drop of G_{1c} when A and B become very different.

A useful way of strengthening the AB interface, if A and B are strongly incompatible, amounts to bringing an AB diblock copolymer at the interface: many studies have been carried out on these decorated systems.

VI
Conclusions

1 Open problems

Long lectures like these tend to be over-optimistic, giving the impression that most physical questions are under control. The reality is different: soft interfaces are far from a happy end. Let me give a few examples:

On the dynamics of wetting: the role of *molecular processes* is not fully appreciated.

(*a*) Following Blake's ideas, they may sometimes be dominant (at large dynamic contact angles). The difficulty is that they are critically dependent on the atomic structure of the surface.

(*b*) When the liquid induces a real chemical reaction on the supporting solid (e.g. a silanation on the OH groups of a silica surface), the exact nature of the driving force is subtle: what fraction of the reaction enthalpy is directly transformed into heat, and what fraction pulls the contact line?*

---

* **Note added in proof:** We now have a much clearer picture of reactive wetting, based on experiments by T. Ondarçuhu and

(*c*) In the case of Aztec pyramids, for instance with one molecular layer spreading out from a thicker region, the description of this layer as a two-dimensional liquid is open to some doubt: we may, in some cases, be dealing with a two-dimensional gas rather than a two-dimensional liquid.

(*d*) The role of surface rugosity is important for these thin layers. It may be that the spreading molecules follow preferentially certain channels (or steps) on the surface: the percolation properties of the channel network may be essential.

(*e*) The action of surfactants is important and not fully understood (although J. F. Joanny did look at some basic features): the transfer of amphiphiles at the contact line from the liquid/air interface to the solid surface may be an important feature (figure 42) and we know very little about it – in particular how do the rates of transfer change when we go from a static line to a moving line?

Even the *hydrodynamics* of wetting and dewetting raises many unsolved questions: for instance, when we observe the growth of a dry patch (a hole) in a liquid film, what is the exact shape of the rim? The simple picture of ref. [1] (with a portion of a circle as the

Dominguez dos Santos (*Phys. Rev. Lett.*, **75**, p. 2976 (1995)). A theory can be found in F. Brochard-Wyart and P. G. de Gennes, *C. R. Acad. Sci.* (Paris), **321 II**, pp. 285–8 (1995).

Conclusions

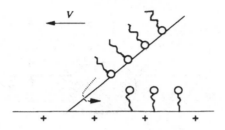

Figure 42. The transfer of amphiphiles at
the contact line from a liquid/air interface
to the solid surface.

profile) is naive, and certainly not valid for large contact angles. (Also the logarithmic cut-offs at both ends
of the rim are not the same, as pointed out by F. Brochard-Wyart; this has to be incorporated to get good
numerical predictions, even for small angles.)

More dramatically, there are cases where no rim is
present! This has been observed by Debregeas *et al.*[†]
on a slightly different problem: opening a hole in a
freely suspended film of a very viscous polymer. But
similar features seem to occur also when the polymer
film is floating on a non-viscous liquid (figure 43
overleaf).

[†] *Phys. Rev. Lett.*, **75**, pp. 3886–9 (1995).

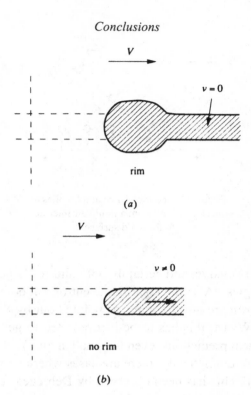

Figure 43. In contrast to soap films (*a*), no rim formation is seen when a polymer film (*b*) is floating on a non-viscous liquid.

Conclusions

We find even more uncertainty if we go to *slippage* of a polymer melt against a solid wall. For walls which have trapped permanently some polymer chains, we do expect a low velocity regime with negligible slip, and a higher velocity regime with a certain critical stress σ^*, and increasing slippage lengths b. But if we go beyond this, we might find other dramatic events, such as complete decohesion between the flowing melt and the fixed chains; or tear-out of the fixed chains.

Also, the surface density of fixed chains plays a non-trivial role. Recently C. Gay and F. Brochard-Wyart have investigated regimes where the fixed chains do not simply add up their contributions, but work in a cooperative fashion. Here, the transition from no slip to slip should require higher velocities: this will hopefully be checked by further experiments. But ultimately, to attack wall effects in extrusion, we must also incorporate both the roughness of the surface and the viscoelasticity of the melt phase. This may take a long time.

Adhesion science is also somewhat immature:

(*a*) For glassy polymers, we can use the Brown model when the material deforms by crazing (like polystyrene or polymethyl metacrylate). But when it responds differently to strong stresses, we are in unknown territory. What happens if fracture is announced by deformation bands? What happens in

101

cross-linked systems? What happens in partly crystalline materials?

(*b*) For soft materials such as rubbers, or liquid tackifiers, I have the (bold) feeling that we understand some of the basic viscoelastic processes (see equation 62). But we are still far from predicting the complete curve $G(V)$ for peeling at a velocity V, when our adhesive has a realistic distribution of relaxation times. And the often proposed proportionality between $G(V)$ and the loss modulus $\mu''(\omega)$ cannot, in general, be justified by theory.

(*c*) We discussed the theoretical role of connector molecules (between a rubber and a solid surface, or between two rubber pieces). But we must mention recent experiments by L. Léger, M. Deruelle and co-workers, where a silica surface is decorated with PDMS connectors, and exposed to a PDMS rubber. These experiments show only a very weak adhesion when the silica is grafted in a (supposedly) clean way: this is in strong disagreement with our model! It may be that the grafted chains do not manage to interdigitate with the rubber network, when we prepare the adhesion junction: either because of some unexplained incompatibility (both partners are PDMS) or because of very long time constants for penetration. But, as of now, we do not understand these experiments.

(*d*) Concerning the tackiness of linear polymers, we are in great need of fundamental experiments:

where the adhesion energy would be measured on really *flat* surfaces (avoiding the complications due to roughness which Creton and Leibler discuss). We especially need data where incubation time and separation time are controlled *separately*.

The *welding problem* (between two blocks of the same polymer) also suffers from many difficulties:

(*a*) The preparation of the interface is delicate. The Lausanne group spent a lot of effort to achieve good contact between the partners at the initial stage. But we have seen that the original distribution of chain ends near the contact plane may be critical: and we have, up to now, no independent probe of this distribution.

(*b*) In the present talk, we discussed only the symmetrical case, where the chain lengths are the same in both blocks $N_A = N_B = N$. The dissymmetric case $N_A \gg N_B$ is quite subtle. E. Kramer and coworkers showed the first basic feature, namely that the short chains migrate relatively fast into the long ones. But there are further complications, because at short times the long portion behaves like a gel, and is able to swell only up to a certain limit. In the extreme case where the short chains are simply monomers ($N_B = 1$) we are still facing interesting questions. When the glass transition of A is high, while B is liquid, if we operate at an intermediate temperature, we often find that the penetration front is not diffusion, but moves at constant velocity. This type II diffusion has been the source of various interpretations. Sarti emphasised the possible role of crazes while A. Kramer and Hui think in terms of viscous flows, with a viscosity which is dramatically dependent on the concentration of the B species. Recently, Rossi, Pincus and I proposed another approach, based on the maximum jump velocity of one B molecule in the A matrix. All these ideas may play a role in favourable conditions, but as of now, we are far from any consensus.

2 Two remarks on style

(*a*) Compared with the giants of quantum physics, we soft-matter theorists look like the dwarfs of German folk tales. These dwarfs were often miners or craft-workers: we, also, are strongly motivated by industrial purposes. We see fundamental problems emerging from practical questions – aquaplaning leads to a new form of dewetting, etc. But there is another, equally important motivation: the challenges of everyday life. Let me take an example, which I owe to the director of the Tefal company. These days, a lot of time is still spent on *ironing*: in a country the size of England, something like ten million people ironing for one hour a week. If, by some intelligent reflection, we can gain, say, six minutes on this hour, a 10% effect, we are saving 10^5 man-hours per day – we are providing a non-negligible improvement for many individuals who come back exhausted from their work. Personally I would feel more proud to achieve this than to solve an elegant formal problem in statistical physics. (Unfortunately, I fail on both counts.)

(*b*) Science is clearly a form of art, with the same invention and the same doubts. There are major differences however: one is the difficulty of communication. An Indian playing his flute in the streets of Bogota invents a new tune: within ten seconds, any passer-by may be stuck by it — possibly for their whole life. But in our trades, a beautiful discovery can be transmitted only to people who have been through a long, specialised education. We must do our best to keep in contact with our fellow citizens, but we often fail.

Incidentally, the artistic professions suffer from many parasites: among others, the art critics or commentators. Fortunately, we do not have the counterpart of art critics in our sciences (although some referees tend to mimic this style ...).

But the analogy between, say, chemists and sculptors, or physicists and printers, is, on the whole, rather close. The great scientists of the early quantum period were producing heroic pictures of the whole universe. After this, there is a natural trend towards a certain form of baroque; or towards artists of small details (I am fond of both). Then, there is the time when the competition of black-and-white photography with painting catalyses a deep change. Exact reproductions of nature become a standard operation. What is discovered, at this stage, is the interest (and difficulty) of extracting a simple vision from a scene, 'an *impression*'. I contend that we have similar trends in soft-matter physics. Simulations and other numerical exercises are the analogue of photography. But what we need most is a simple impressionist vision of complex phenomena, ignoring many details – actually, in many cases, operating only at the level of scaling laws.

Thus, I tend to compare our community of soft-matter theorists to the amateur painters of a hundred years ago – spending their Sunday afternoons in the park, and capturing a few simple scenes – involving their friends, their children, and those they love. I see no better style.

Appendix
Drag on a tethered chain moving in a
polymer melt

A chain of N monomers is attached to a small colloidal particle, and is pulled (at a velocity V) inside a polymer melt (chemically identical, with P monomers per chain). The main parameter for this problem is the number $X(V)$ of P chains entangled with the N chain. Earlier estimates of X are criticised in this appendix, which is based on work by A. Ajdari, F. Brochard-Wyart, C. Gay, and J. L. Viovy (1995), and a new form is proposed: at large N ($N > N_e^2$), we are led to a 'Stokes' regime, $X = N^{1/2}$, while at smaller N ($N < N_e^2$), we find a 'Rouse' regime, $X = N / N_e$ (where N_e is the number of monomers per entanglement).

The motion of a long tethered chain (N monomers) inside a polymer melt (P) is special: the N chain cannot reptate inside the P matrix. This occurs in star polymers, and also in two recent experimental situations (figure 2):

(a) The N chain is grafted to a colloidal particle (of

107

size smaller than the coil radius R_N of the N chain). The particle can be driven by sedimentation or by optical tweezers.

(*b*) The N chain is grafted on a flat wall, and the P melt flows tangentially to the wall (figure 15). (In all that follows, we assume that the grafting density is very small: no coupling between different N chains.)

Problem (*b*) was first considered theoretically (for the low V limit) in reference [30]. The starting point is that a certain number $X(V)$ of P chains are entangled with the N chain. The resulting friction is estimated as follows:

Assume that the N chain has moved by a distance D^* equal to the diameter of an Edwards tube [4]. $D^* = N_e^{1/2} a$, where N_e is the number of monomers per entanglement, and a the monomer size. (We take $N_e < N \leq P$.) To allow for this motion of the N chain, each P chain entangled with N must move along its own tube (of length L_t) by something like L_t. Thus the sliding velocity V_s of this P chain is not the translational velocity V, but is much larger:

$$V_s \cong V \frac{L_t}{D^*} = V \frac{P}{N_e} \tag{A1}$$

The dissipation $T\dot{S}$ due to the motion of the tethered chain corresponds to $X(P)$ chains moving at velocity V_s in the ambient melt:

$$T\dot{S} = X\zeta_1 P V_s^2 = fV \qquad (A2)$$

where $\zeta_1 P$ is the tube friction coefficient of one P chain, and f the drag force. Comparing the two expressions of $T\dot{S}$ we get:

$$f = VX(V)a\,\eta_p \qquad (A3)$$

where $\eta_p = \zeta_1 a^{-1} P^3 N_e^{-2}$ is the reptation viscosity of the (P) melt.

The crucial question is thus to find $X(V)$. Even in the simplest $(V \to 0)$ limit, where the N chain is an unperturbed coil, this problem is difficult, and different answers have been proposed at different times [14]. We reanalyse the problem here, using what we call the binary entanglement model. We also compare this with a 'collective' entanglement model. Finally, we extend our ideas to higher velocities.

At low velocities, the N chain is an unperturbed coil, of size $R_N \cong N^{1/2}a$ and volume R_N^3. It experiences on average an entanglement every N_e monomers. Each P chain intersecting this volume uses ~N monomers in this region. Thus the number of P chains which overlap with the N coil is $R_N^3 / Na^3 = N^{1/2}$.

In reference [30] we simply assumed that all the P chains are entangled with the N chain, that is, $X(V \to 0) = N^{1/2}$.

The Edwards tube surrounding the N chain (figure

19) is a sequence of N / N_e blobs with diameter D^* and total volume:

$$\Omega = N / N_e (D^*)^3 \qquad (A4)$$

One of the P chains intersecting the volume R_N^3 has $N\Omega / R_N^3$ monomers inside the tube. The number of blobs visited by the P chain is thus:

$$b = \left(\frac{N}{N_e} \right) \left(\frac{\Omega}{R_N^3} \right) = \left(\frac{N}{N_e} \right)^{1/2} \qquad (A5)$$

Inside one blob, $N_e^{1/2}$ chains coexist (including the N chain). In our binary entanglement model, we assume that a constraint is associated with a *pair* of chains inside the blob. The total number of pairs is $\frac{1}{2}(N_e^{1/2})^2 \sim N_e$. Thus, the probability that any given pair of chains inside the volume do entangle, is only $N_e^{-1/2}$. The number c of constraints between one (P) chain and the (N) chain is then:

$$c = bN_e^{-1/2} = \frac{N^{1/2}}{N_e} \qquad (A6)$$

We are thus led to distinguish two very different regimes:

(1) $N > N_e^2$. In this case c is larger than unity: all the $N^{1/2}$ (P) chains which intersect the coil do entangle

with the (N). Thus the simple guess of ref. [30] is confirmed:

$$X = N^{1/2} \tag{A7}$$

We call this the Stokes regime, because the friction force (equation A3) has the scaling form corresponding to a Stokes sphere (radius $N^{1/2}a$) inside a liquid of viscosity η_p.

(2) $N < N_e^2$. In this case c is smaller than unity, and we cannot use equation (A5). When $c \ll 1$, we may say that the probability of entanglement between one (P) chain (intersecting the coil) and the (N) chain is c. Thus:

$$X = N^{1/2}c = \frac{N}{N_e} \tag{A8}$$

The friction experienced by the tethered chain is then *linear in N*. Although we deal with an entangled system, the (N) chain is thus described by the so-called Rouse model [31], but the Rouse friction coefficient is proportional to the melt viscosity.

We now describe an opposite limit, where one entanglement site (a blob of diameter D^*) is pictured as a very complex knot, involving $N_e^{1/2}$ chains; the knot is such that eliminating one chain from it is enough to remove the constraint. Then the number of constraints released if one P chain moves out of the volume R_N^3 is b, and is larger than unity. All the $N^{1/2}$ 'P chains' are thus coupled to the N chain.

In this model, it would be enough to select a subset of $(N / N_e)^{1/2}$ 'P chains' and move them out, to relax the N chain: since $(N / N_e)^{1/2} b = N / N_e$ is the total number of constraints to be removed. This remark leads to the prediction of reference [14]. However, we do not think that this approach is realistic. The N chain, when it moves, has no way of selecting a subset of releasing chains: it drags all of them.

Thus we are led to say that, in the collective entanglement model, $X = N^{1/2}$, and the Stokes model holds for all values of N.

Under strong flows, and in the simplest picture [13], the N chains become elongated into a cigar shape, with diameter D and length $L = R_N^2 / D$. A more sophisticated description has been constructed [14], but is essentially equivalent in practice. In the binary entanglement model, we have to distinguish two regimes:

(*a*) *Partial striction*: $R_N > D > D^*$. Here, a simple repetition of our discussion in section II gives:

$$c = \frac{D}{aN_e} \qquad (A9)$$

If $N < N_e^2$, we always stay in the Rouse regime ($X = N / N_e$).

If $N > N_e^2$, we find a crossover from strong

coupling to Rouse upon increasing the velocity (decreasing D).

(*b*) *Marginal regime*: Here the cigar diameter D becomes comparable to D^*, and the number of entanglements realised by the N chain can become smaller than N / N_e. The marginal value of X ($X = X^*$) is in fact fixed by the force balance: the stretching force required to reach D^* is:

$$\frac{kT}{D^*} = X^*(V)a\eta_p V \qquad (A10)$$

and thus X^* is inversely proportional to the velocity. It may be checked that for $V = V^*$ (the onset velocity for the marginal regime) $X^* = N / N_e$ as expected in the Rouse regime and:

$$V^* = \frac{kTN_e^{1/2}}{N\eta_p a^2} \qquad (A11)$$

Remarks

(*a*) The results of the binary entanglement model can be summarised as follows (for the low velocity limit): the number of entangled chains is either $N^{1/2}$ (the number of ambient chains intersecting the mushroom) or N / N_e (the number of constraints acting on the N

chain). Each of them is an upper bound for X, and thus X is the smallest of the two.

(*b*) In the collective entanglement model, we are led to $X = N^{1/2}$. But we do not think that the collective model is fully realistic: complex knots may play a role, but may not dominate the behaviour. Thus we tend to stick to the binary model.

(*c*) It is instructive to discuss the whole distribution function $p(n)$ for the number of entanglements between one given P chain (intersecting the mushroom) and the N chain. Using mean field arguments, one arrives at a Poisson distribution. This gives:

$$X = N^{1/2}[1 - p(0)]$$
$$= N^{1/2}\left[1 - \exp\left(\frac{-N^{1/2}}{N_e}\right)\right] \qquad \text{(A12)}$$

Equation (A12) is a useful interpolation between equations (A7) and (A8).

References

[1]   F. Brochard-Wyart, P. G. de Gennes, *Adv. Colloid Interface Sci.*, **39**, 1 (1992)

[2]   R. G. Cox, *J. Fluid Mech.*, **168**, p. 169 (1966)

[3]   P. G. de Gennes, X. Hua, P. Levinson, *J. Fluid Mech.*, **212**, 55 (1990)

[4]   C. Ngan, E. Dussan, *J. Fluid Mech.*, **118**, 27 (1982)

[5]   P. G. de Gennes, *Rev. Mod. Phys.*, **57**, 827 (1985)

[6]   F. Heslot, N. Fraysse, A. M. Cazabat, *Nature*, **338**, 1289 (1989)

[7]   P. G. de Gennes, A. M. Cazabat, *C.R. Acad. Sci.* (Paris), **310 II**, 1601 (1990)

[8]   N. Fraysse, A. M. Cazabat, M. Cazabat, *C.R. Acad. Sci.*, **314 II**, 1025 (1992)

[9]   P. G. de Gennes, *C.R. Acad. Sci.* (Paris), **288 B**, 219 (1979)

[10]   F. Brochard-Wyart, P. G. de Gennes, H. Hervet, C. Redon, 'Wetting and slippage of polymer melts on semi ideal surfaces', *Langmuir*, **10**, 1566 (1994)

[11]   R. Burton, M. Folkes, N. Narh, A. Keller, *J. Mat. Sci.*, **18**, 315 (1987)

[12]   K. Miegler, H. Hervet, L. Léger, *Phys. Rev. Lett.*, **70**, 287 (1993)

References

[13] F. Brochard-Wyart, P. G. de Gennes, *Langmuir*, **8**, 3033 (1992)

[14] A. Ajdari, F. Brochard-Wyart, P. G. de Gennes, L. Leibler, J. L. Viovy, M. Rubinstein, *Physica A* **204**, 17 (1994)

[15] See, for instance, J. F. Knott, *Fundamentals of Fracture Mechanics*, Wiley (1973)

[16] E. Raphaël, P. G. de Gennes, *J. Phys. Chem.*, **96**, 4002 (1992)

[17] P. G. de Gennes, *Macromolecules*, **13**, 1069 (1980)

[18] H. Brown, *Macromolecules*, **24**, 2752 (1991)

[19] H. Jud, H. Kausch, J. Williams, *J. Mat. Sci.*, **16**, 204 (1981)

[20] H. Kausch, D. Delacretaz, *Proc. of the IBM Symposium on Polymers*, Lech (1990)

[21] D. Legrand, G. Gaines, *J. Colloïd Interface Sci.*, **31**, 162 (1969)

[22] A. Harihakan, S. Kumar, T. Russell, *Macromolecules*, **23**, 3584 (1980)

[23] P. G. de Gennes, *C.R. Acad. Sci.* (Paris), **307 II**, 1841 [1988]

[24] P. G. de Gennes, *J. Chem. Phys.*, **55**, 572 (1971)

[25] P. G. de Gennes, *C.R. Acad. Sci.* (Paris), **B 291**, 219 (1980)

[26] S. Prager, M. Tirrell, *J. Chem. Phys.*, **75**, 5194 (1981)

[27] R. P. Wool, R. O'Connor, *J. App. Phys.*, **52**, 5953 (1981)

References

[28] G. Reiter, U. Steiner, *J. Phys.* (Paris), 659 (1991)

[29] Y. Iyengar, D. Erickson, *J. App. Polymer Sci.*, **11**, 2311 (1967)

[30] F. Brochard-Wyart, P. G. de Gennes, P. Pincus, *C.R. Acad. Sci.*, **314 II**, 873 (1992)

[31] For a recent presentation of the Rouse model, see P. G. de Gennes, *Introduction to Polymer Dynamics*, Cambridge U. Press (1990)